All About Moon Bases-And Our Plans to Return to the Moon

NASA has plans to land on the Moon again in 2024 and then build a Moon base following that.

What is the history of plans to build a base on the Moon and what are all the issues involved?

Where should the base be built and what will be do once we have long term habitation on the Moon?

There are lots of ideas and lots of issues to consider.

My hope is that the reader will learn about all of the issues important to building a Moon Base to better understand what will be required.

You might even want to become a Moon Base resident!

All About Moon Bases-And Our Plans to Return to the Moon

All About Moon Bases-And Our Plans to Return to the Moon

All About Moon Bases-And Our Plans to Return to the Moon

Other books by Martin K. Ettington

Spiritual and Metaphysics Books:
Prophecy: A History and How to
    Guide
God Like Powers and Abilities
Enlightenment for Newbies
Removing Illusions to Find True
    Happiness
Using the Scientific Method to
    Study the Paranormal
A Compendium of Metaphysics
    and How to Guides (Six
    books together in one
    volume)
Love From the Heart
The Enlightenment Experience
Learn Your Soul's Purpose
Pursuing Enlightenment
A Modern Man's Search for Truth

Longevity & Immortality:
Physical Immortality: A History and
    How to Guide
The Commentaries of Living
    Immortals
Records of Extremely Long Lived
    Persons
Enlightenment and Immortality
Longevity Improvements from
    Science
The 10 Principles of Personal
    Longevity
Telomeres & Longevity
The Diets and Lifestyles of the
Worlds Oldest Peoples
The Longevity Six Books Bundle

Science Fiction:
Out of This Universe
Personal Freedom-Parts 1 & 2
The Psychic Soldier Series:
    Book 1-Himalayan Journey
    Book 2-A Soldier is Born
    Book 3-Fighting For Right

    Book 4-Earth Protector
    Book 5-War on the Astral Plane
The Immortality Sci Fi Bundle

The God Like Powers Series:
Human Invisibility
Invulnerability and Shielding
Teleportation
Psychokinesis
Our Energy Body, Auras, and
Thoughtforms
The God Like Powers Series—
Volume 1 Compilation

The Yoga Discovery Series:
Yoga-An Ancient Art Form
Hatha Yoga-Helping you Live
    Better
Raja Yoga-Through the Ages
The Yoga Discovery Package

Business Books:
Creating, Publishing, & Marketing
    Practitioner Ebooks
Building a Successful Longevity
    Coaching Business
Why Become a Coach?
The Professional Coaching
    Success Trilogy

Science and Technology
Future Predictions By and
    Engineer & Seer
The Unusual Science &
    Technology Bundle
The Real Atlantis-In the Eye of the
    Sahara
Ancient & Prehistoric Civilizations
Ancient and Prehistoric
    Civilizations-Book Two
Are Cryptozoological Animals Real
    or Imaginary?

All About Moon Bases-And Our Plans to Return to the Moon

<u>Aliens and Space</u>

Aliens and Secret Technology
Aliens Are Already Among Us
Designing and Building Space Colonies
Humanity and the Universe

<u>The Longevity Training Series</u>

(A transcription of the online Multimedia Longevity Coaching Training Program)

The Personal Longevity Training Series-Book1-Long Lived Persons
The Personal Longevity Training Series-Book2-Your Soul's Purpose
The Personal Longevity Training Series-Book3-Enable Your Life Urge
The Personal Longevity Training Series-Book4-Your Spiritual Connection
The Personal Longevity Training Series-Book5-Having Love in Your Heart
The Personal Longevity Training Series-Book6-Energy Body Health
The Personal Longevity Training Series-Book7-The Science of Longevity
The Personal Longevity Training Series-Book8-Physical Body Health
The Personal Longevity Training Series-Book9-Avoiding Accidents
The Personal Longevity Training Series-Book10-Implementing These Principles

The Personal Longevity Training Series-Books One Thru Ten

These books are all available in digital and printed formats from my website and on Amazon, Barnes & Noble, and Apple ITunes
Website: http://mkettingtonbooks.com

All About Moon Bases-And Our Plans to Return to the Moon

<u>Signup for our Mailing List to get the following</u>

1) A discount coupon for 25% discount on all books on our site

2) Occasional Notices of new books available

3) Occasional Email on other offerings of ours (Monthly)

Go to this link to sign-up:

http://personal-longevity.com/mkebooks/emailsignup/

And click this link to get the FREE 102 page Ebook titled "Secrets of Many Things"

If you have any questions about this book or other subjects please contact the Author at:

mke@mkettingtonbooks.com

All About Moon Bases-And Our Plans to Return to the Moon

All About Moon Bases-And Our Plans to Return to the Moon

Table Of Contents

1.0 Introduction

This is a book about Moon bases and living on the Moon. I like to write about popular topics which interest me and the Moon is one of them.

Maybe it has to do with my growing up in the 1960s at the height of the Apollo Program. I remember walking to school at about seven years old and the kids in the group were all talking about the Mercury Astronauts and how cool they were. Then I followed all of the succeeding Gemini and Apollo missions. It was an incredible experience to watch the Apollo 11 landing and first Moonwalk at Boy Scout camp with us all gathered in the dining hall.

Later on I worked for Hewlett Packard at the NASA Human Spaceflight Center in Houston for a couple of years in the mid nineteen eighties. I also met many of the Astronauts including July Resnick who was killed on the Challenger Shuttle when it blew up at launch. Watching President Reagans' speech at the Johnson Space Center as a memorial to them was also quite the experience. One of my friends who took me up in his Pitt Special acrobatic plane a couple of times later became an Astronaut on the MIR, Space Shuttle, and Space Station. I got my Pilots license at Laporte Airport where many of the off duty Astronauts flew acrobatic planes for fun.

I also applied to the Astronaut Corp but being a civilian and not having any degrees beyond my B.S. in Engineering Science filtered me out. Never the less, I've always been a fan of the space program. I still follow all the details I can to this day. In the last couple of years I wrote a comprehensive guide to Space Colonies titled "Designing and Building Space Colonies-A Blueprint for the Future". In that book I covered a lot of the issues about living in space and construction of space colonies.

With the birth of the Artemus Program and plans to land man on the Moon again in 2024, I thought this would be a good time to go into the history of Moon landings and proposed Moon colonies and maybe make a few of my own suggestions too.

I hope you enjoy this journey into our near future!

2.0 Earth and Lunar Environments

The Moon and the Earth are very different environments. It is important to emphasize the differences in the living conditions of these two spheres to better understand what will have to be built into structures to live safely on the Moon.

Gravity

The Moon's mass is about 1/80th (1.2%) of the Earth's mass, so the Moon's gravity is much less than the Earth's gravity; specifically, the Moon's gravity is 1/6th (16.7%) of the Earth's gravity. Or, stated another way, the Moon's gravity is 5/6 (83.3%) LESS than the Earth's.

This means that you will weigh much less on the Moon but the inertia of objects in motion will remain the same. So you can lift a lot, but will still have trouble stopping a large object in motion.

Atmosphere

On Earth at sea level the atmospheric pressure is 14.7 pounds per square inch. The Moon doesn't have an atmosphere and is basically a vacuum. This means all structures will need to be airtight and have airlocks for going in and out. Of course you will need to wear a spacesuit on the surface on the Moon

Radiation

The surface of the Moon is baldly exposed to cosmic rays and solar flares, and some of that radiation is very hard to stop with shielding. Furthermore, when cosmic rays hit the ground, they produce a dangerous spray of secondary particles right at your feet. All this radiation penetrating human flesh can damage DNA, boosting the risk of cancer and other maladies. Spacesuits and buildings need to be able to protect humans from this radioactive environment.

Of course the atmosphere on Earth protects us from these deadly rays.

Temperature

Daytime on one side of the Moon lasts about 13 and a half days, followed by 13 and a half nights of darkness. When sunlight hits the Moon's surface, the temperature can reach 260 degrees Fahrenheit (127 degrees Celsius). When the sun goes down, temperatures can dip to minus 280 F (minus 173 C)

Earth's temperature varies from the extremes of 132 degree Fahrenheit to minus 126.6 degrees Fahrenheit.

These temperature extremes also mean that your spacesuit needs to be designed to withstand large temperature shifts.

Water

The Earth is a watery planet which is covered 71 percent by water in Oceans and lakes.

We used to think that there was no water on the Moon. Now we have satellites which show there should be water ice at the Poles, especially the South Pole. The question is can we mine it and purify it?

All About Moon Bases-And Our Plans to Return to the Moon

3.0 Requirements for Building Moon Bases

Radiation Protection

The walls of any base structure will need to protect the inhabitants from normal ionizing radiation or even from much larger occasional solar flare radiation. This is a good reason to build the base underground or build a structure then bury it.

Power Sources

Power can be nuclear, solar, or some type of generator running hydrogen from water split into hydrogen and helium. One of these sources might be a fuel cell. There might need to be battery backup too. If the base is built in a South Pole crater like Shakelton crater then solar will be a problem since there is no Sun in the crater. One possibility is putting solar panels on the crater rim in the sunlight and running cables back to the base.

Atmosphere

Any shelter will need to provide an atmosphere where the residents can work in a shirtsleeve environment. There will also need to be safety protections in case of a puncture of the structure(s). This probably means having airlocks and/or building the shelter underground or covered by soil to reduce the possibility of loss of air.

Being able to get oxygen from cracking water into hydrogen and oxygen will help provide availability of this critical gas to breathe.

Water

Water will need to be provided and recycled to clean it. The ability to mine water ice, purify it, and melt it will be a great boon to the residents. It would be very costly to ship all the water needed from Earth's gravity well.

Temperature Control

The surface temperature varies from 260 degrees plus Fahrenheit to 280 minus degrees Fahrenheit needs to be insulated against in both shelters and spacesuits. This has already been addressed in Apollo spacesuit design and buildings would need to be built to these standards.

Vehicles for the surface

Vehicles moving around the Moon need to have a much longer range than during the Apollo missions and maybe also an enclosed cabin so travelers can get out of their suits and rest in a shirtsleeve environment.

All About Moon Bases-And Our Plans to Return to the Moon

Communications

If the lunar base is located at the South Pole in Shakelton Crater then it will not be in line of site with the Earth. Also, it will need to communicate with the Lunar Gateway circling the Moon. This probably necessitates having satellites in orbit around the Moon to connect everyone together.

Growing Food

It costs a lot of money to ship supplies to the Moon from Earth for a crew to stay on the Moon for extended periods. It would be much more cost effective to allow Moon workers to grow at least some of their own food hydroponically. This would also provide some fresh fun for the persons growing the food.

Sewage

Having some way to recycle or process sewage would also make sense. It might even be useful to help grow some crops.

All About Moon Bases-And Our Plans to Return to the Moon

4.0 Moon Bases in Science Fiction

There are many science fiction novels about the Moon but I've picked these five which in many ways have affected our thinking and plans for visiting the Moon.

4.1 From the Earth to the Moon by Jules Verne

The first modern story about visiting the Moon was Jules Verne's book "From the Earth to the Moon" written in 1865. It was followed by his sequel "Around the Moon"

Jules Verne was very prescient about several things such as:

a) A Moon Launch from Florida
b) Three men in the Capsule
c) About weightlessness and the vacuum of space.

This story seems to have led to the modern interest in travel to the Moon.

Here is a summary of the second book "Around the Moon"

> Having been fired out of the giant Columbiad space gun, the Baltimore Gun Club's bullet-shaped projectile, along with its three passengers, Barbicane, Nicholl and Michael Ardan, begins the five-day trip to the Moon. A few minutes into the journey, a small, bright asteroid passes within a few hundred yards of them, but does not collide with the projectile. The asteroid had been captured by the Earth's gravity and had become a second moon.

> The three travelers undergo a series of adventures and misadventures during the rest of the journey, including disposing of the body of a dog out a window, suffering intoxication by gases, and making calculations leading them, briefly, to believe that they are to fall back to Earth. During the latter part of the voyage, it becomes apparent that the gravitational force of their earlier encounter with the asteroid has caused the projectile to deviate from its course.

> The projectile enters lunar orbit, rather than landing on the Moon as originally planned. Barbicane, Ardan and Nicholl begin geographical observations with opera glasses. The projectile then dips over the northern hemisphere of the Moon, into the darkness of its shadow. It is plunged into extreme cold, before emerging into the light and heat again. They then begin to approach the Moon's southern hemisphere. From the safety of their projectile, they gain spectacular views of Tycho, one of the greatest of all craters on the Moon. The three men

discuss the possibility of life on the Moon, and conclude that it is barren. The projectile begins to move away from the Moon, towards the 'dead point' (the place at which the gravitational attraction of the Moon and Earth becomes equal). Michel Ardan hits upon the idea of using the rockets fixed to the bottom of the projectile (which they were originally going to use to deaden the shock of landing) to propel the projectile towards the Moon and hopefully cause it to fall onto it, thereby achieving their mission.

When the projectile reaches the point of neutral attraction, the rockets are fired, but it is too late. The projectile begins a fall onto the Earth from a distance of 260,000 kilometres (160,000 mi), and it is to strike the Earth at a speed of 185,400 km/h (115,200 mph), the same speed at which it left the mouth of the Columbiad. All hope seems lost for Barbicane, Nicholl and Ardan. Four days later, the crew of a US Navy vessel, Susquehanna, spots a bright meteor fall from the sky into the sea. This turns out to be the returning projectile, and the three

4.2 The First Man in the Moon by H.G. Wells

Originally published in 1901, the novel has been re-released 100 years later. The publishers understand that presentation is just as important as the overall story and the book greets the reader with a stunning cover illustration by Chris Moore, an artist trained at the Royal College of Art, who proves more than capable of capturing the essence of this adventure.

Set in England at the beginning of the 20th century, average industrialist Bedford finds himself entwined in the machinations of Cavor, an eccentric genius who has developed Cavorite, a substance that negates the pull of gravity. The two men construct a vessel called the Sphere which hurls them to the Moon. But the adventurers have very different agendas. Cavor hopes to discover a utopian society he imagines living on the planet, while Bedford is purely interested in the monetary gain the trip represents (after all, everyone knows there's gold on the Moon). Once they arrive, they stumble upon the world of the Selenites, insect-like, biologically engineered aliens living beneath the surface of the Moon in dark, cavernous, technologically-astounding cities. Then things go drastically wrong...

Wells' envisioning of Earth's satellite is fascinating in its accuracy; a barren planet with a thin (yet breathable) atmosphere, a freezing night and very little gravity. However, when the sun rose, Wells imagined forests of trees and plants exploding to life, having a mere Moon-day (which is like an Earth week) to grow, germinate and seed before the cold of the night withers them. Wells saw the possibility that the Moon itself would be full of catacombs, tunnels and internal seas. The Selenite society (although Cavor humorously refers to them as "Moonies") would exist beneath the surface like an ant colony. The images he creates are briefly seen by Bedford and somewhat described later by Cavor, which Wells has cleverly done to leave the reader's imagination to paint its own picture of this underworld

4.3 The Moon is a Harsh Mistress By Robert Heinlein

This is my favorite Science Fiction Novel of all time.At the time of the story, 2075, the Moon (Luna) is used as a penal colony by Earth's government, with the inhabitants living in underground cities. Most inhabitants (called "Loonies") are criminals, political exiles, or their descendants. The total population is about three million, with men outnumbering women two to one, so that polyandry is the norm. Although Earth's Protector of the Lunar Colonies (called the "Warden") holds power, in practice, little intervention exists in the loose lunar society.

Due to the low surface gravity of the Moon, Loonies who stay longer than a few months undergo "irreversible physiological changes and can never again live in comfort and health in a gravitational field six times greater than that to which their bodies have become adjusted".

HOLMES IV ("High-Optional, Logical, Multi-Evaluating Supervisor, Mark IV") is the Lunar Authority's master computer, having almost total control of Luna's machinery on the grounds that a single computer is cheaper than (though not as safe as) multiple independent systems.

The story is narrated by Manuel Garcia "Mannie" O'Kelly-Davis, a computer technician who discovers that HOLMES IV has achieved self-awareness and has developed a sense of humor. Mannie names it "Mike" after Mycroft Holmes, brother of Sherlock Holmes, and they become friends.

The main story is about how moon residents conduct a successful revolution to get their independence from Earth. This story also includes a linear accelerator used to launch payloads into orbit and even to bombard the Earth with rocks.

4.4 The 2001 Movie Moon Base by Arthur C. Clarke

(Movie by Stanley Kubrick)

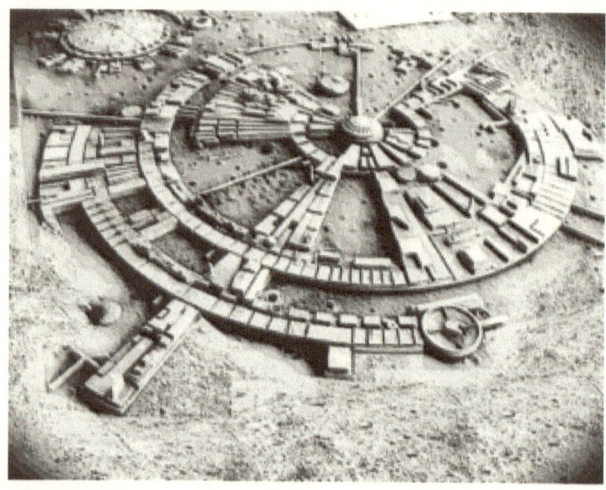

2001 was a movie based on the book 2001 by Arthur C. Clarke. In this book a monolith is found on the Moon emitting a strong signal which another spaceship eventually traces to Jupiter and finds a super race of aliens.

As part of the movie there is vivid trip to a base underground on the Moon which is also shown in this reddish picture below. This picture is part of a sequence where the Moon landing vehicle is descending on an elevator into the underground base.

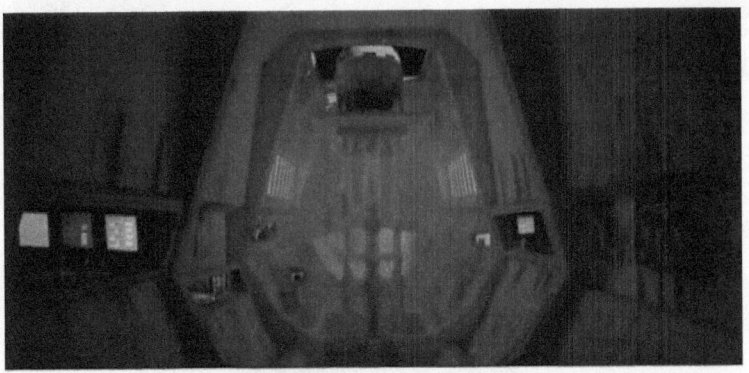

It is a very realistic portrayal of a base underground on the Moon.

4.5 Artemis By Andy Weir

Andy Weir is the Author of an incredible story about a base on Mars which was made into a movie. This story is his take on a Moon Base:

> "Artemis" itself is a five-dome Moon base, servicing a little heavy industry and rather more tourism. Jazz, our heroine, is a sparky young woman who (while her observant Muslim father tut-tuts) gets drunk, has sex and generally tries to have a good time. It's a struggle, though: good times are expensive on the Moon, and despite supplementing her job – she is a porter – with some judicious smuggling, Jazz is always short of money. She lives in a coffin-sized apartment, shares communal washing facilities and eats the cheapest algae-grown gunk. Poverty persuades her to take on a criminal commission: a little light sabotage on the lunar surface. Naturally, things don't go smoothly: she botches the sabotage, her employer gets

murdered, and an assassin is coming after her. The Moon has become a battleground for organized crime over a MacGuffin, in this case a new tech that could revolutionize Earth's entire communication system.

All About Moon Bases-And Our Plans to Return to the Moon

5.0 Early Moon Base Design Proposals

Man has been thinking about visiting the Moon since ancient times. The idea of a base and living on the Moon is much more recent and as far as I can tell goes back to about 1959 and Project Horizon.

The above drawing shows a lunar base for six to twelve people, built into an inflatable spherical habitat.

Proportions of interior volume devoted to different systems equipment is relatively accurate. The heaviest equipment such as for environmental control, and areas in which the crew spends the most time, such as their personal sleep quarters are lowest in the habitat. Work areas for lunar sample analysis, for hydroponics, and even for small animals are located in the middle areas. The top deck in this view is a running track on which the sloped surface permits the crew member to use centripetal force rather than gravity to permit running in 1/6 G. Concept: NASA (1989)

5.1 Project Horizon

Project Horizon was a 1959 study to determine the feasibility of constructing a scientific / military base on the Moon, at a time when the U.S. Department of the Army, Department of the Navy, and Department of the Air Force had total responsibility for U.S. space program plans. On June 8, 1959, a group at the Army Ballistic Missile Agency (ABMA) produced for the Army a report titled Project Horizon, A U.S. Army Study for the Establishment of a Lunar Military Outpost. The project proposal states the requirements as:

> *The lunar outpost is required to develop and protect potential United States interests on the Moon; to develop techniques in Moon-based surveillance of the earth and space, in communications relay, and in operations on the surface of the Moon; to serve as a base for exploration of the Moon, for further exploration into space and for military operations on the Moon if required; and to support scientific investigations on the Moon.*

The permanent outpost was predicted to be required for national security "as soon as possible", and to cost $6 billion. The projected operational date with twelve soldiers was December 1966.

Horizon never progressed past the feasibility stage, being rejected by President Dwight Eisenhower when primary

responsibility for America's space program was transferred
to the civilian agency NASA

5.2 Subsurface Moon bases

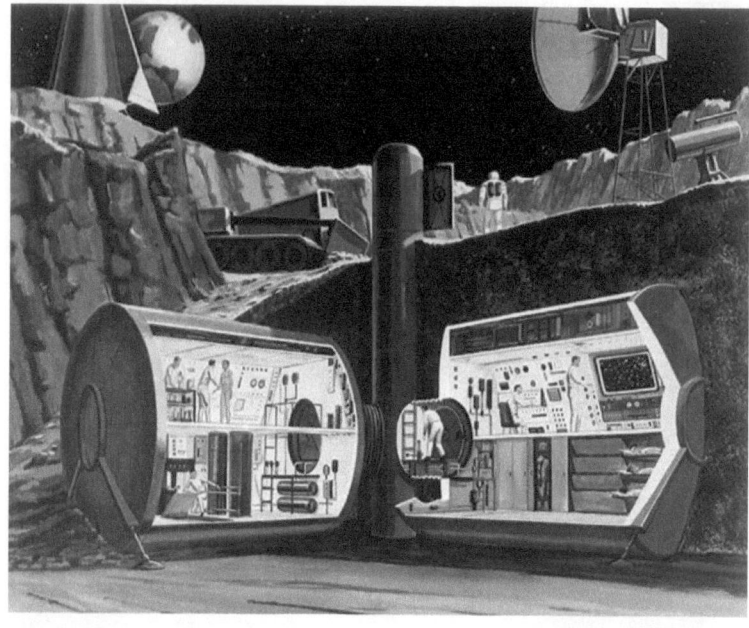

Building a subsurface Moon base was a concept in the early 1960s. Some suggest building the lunar colony underground, which would give protection from radiation and micrometeoroids. This would also greatly reduce the risk of air leakage, as the colony would be fully sealed from the outside except for a few exits to the surface.

The construction of an underground base would probably be more complex; one of the first machines from Earth might be a remote-controlled excavating machine. Once created, some sort of hardening would be necessary to avoid collapse, possibly a spray-on concrete-like substance made from available materials. A more porous insulating material also made in-situ could then be applied. Rowley & Neudecker have suggested "melt-as-you-go" machines that would leave glassy internal surfaces. Mining methods such as the room and pillar might also be used. Inflatable self-sealing fabric habitats might then be put in place to retain air. Eventually an underground city can be constructed. Farms set up underground would need artificial sunlight. As an alternative to excavating, a lava tube could be covered and insulated, thus solving the problem of radiation exposure. An alternative solution is studied in Europe by students to excavate a habitat in the ice-filled craters of the Moon.

5.3 The LESA Moon Base

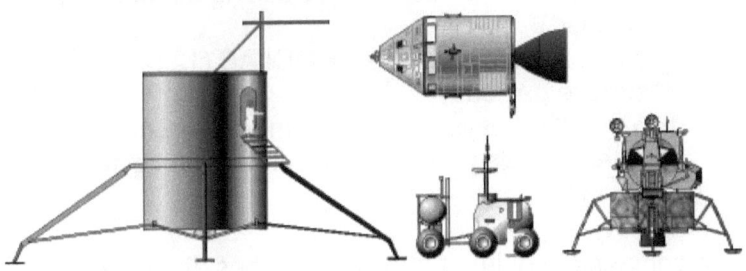

LESA would use a new Lunar Landing Vehicle to land payloads of from10,500 kg to 25,000 kg on the lunar surface with a single Saturn V launch. Extended CSM and LM Taxi hardware derived from the basic Apollo program would allow crews to be rotated to the ever-expanding, and eventually permanent lunar base. A nuclear reactor would provide power.

Evolution to a lunar base would go from the basic Apollo hardware to AES (Apollo Extension Systems) to ALSS (Apollo Logistics Support System using the LEM Truck), and then ultimately to LESA (Lunar Exploration System for Apollo). Modules developed for ALSS or LEM Truck could be used in LESA systems for commonality and to reduce development costs. The end result would be ever-expanding permanent stations on the Moon.

A typical vision of post-Apollo lunar exploration envisioned the following phases:

- 2 men/2 days - Apollo
- 2 men/14 days - AES - LEM Shelter (2050 kg surface payload - LEM Shelter)
- 2 men/14 to 30 days - ALSS with shelter or MOLAB (4100 kg surface payload)
- 3 men/90 days - LESA I (10,500 kg surface payload)
- 3 men/90 days - LESA I + MOLAB (12,500 kg surface payload)
- 6 men/180 days - LESA II with shelter and extended range roving vehicle (25,000 kg surface payload)

In a comparison of lunar base approaches, the basic Apollo hardware scenario for thorough exploration of a single location would consist of a single manned lunar reconnaissance landing of the selected base site, followed by six Apollo launches over the next six quarters - total, 14 man-days on the Moon for 7 Saturn V launches. The AES or ALSS approach would follow the single reconnaissance flight by three pairs of cargo landings and manned landings, resulting in a total of 86 man-days on the Moon for the same number of Saturn V launches. The LESA approach, with a cargo lander followed by two manned landings in sequence to the same large shelter and rover, would allow 542 man-days on the Moon. ALSS development would cost around $500 million, and LESA cost $1.45 billion. In terms of cost per man-day on the Moon, either approach would pay off on the very first mission.

A diagram of the LESA shelter is below:

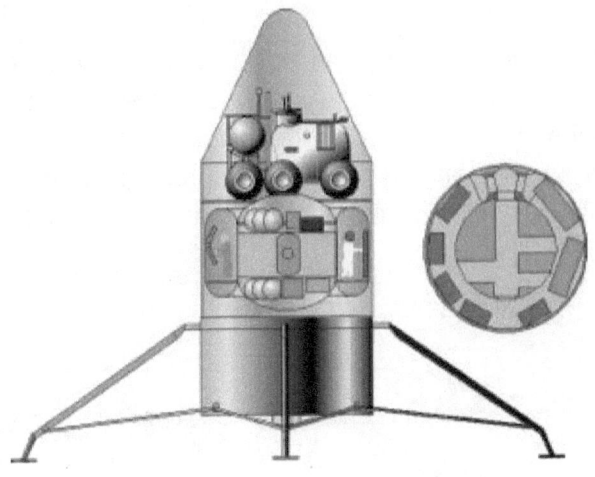

5.4 Underground Base Plan

In 1969 the lunar colony concept (picture below) was developed to encompass a lunar base buried under lunar soil (Johnson, 1969).The sequence of events thought to be possible was a landing in 1969, resources development in 1973-75, a scientific station in 1975, and the lunar colony by 1978.

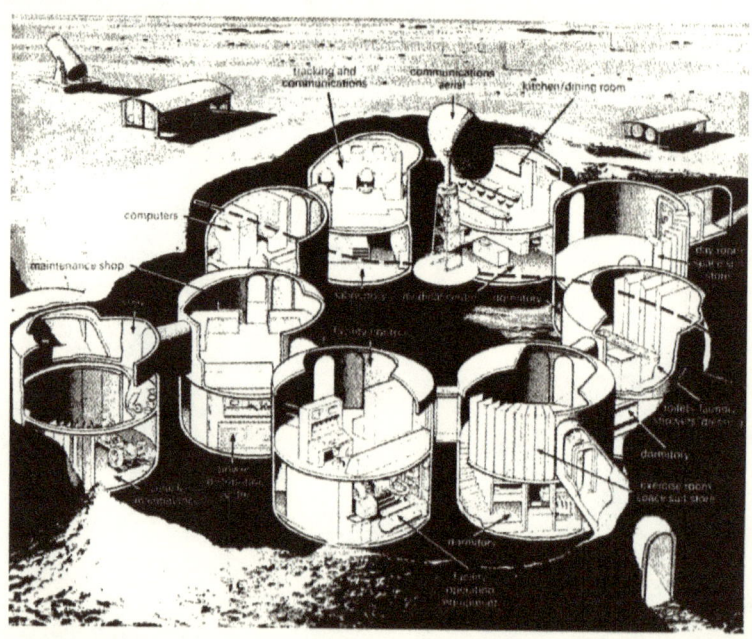

6.0 The First Moon Landings & Recent Missions

Without going into all of the history of the United States manned Moon landings in from 1969 to 1972, it is useful to think about what we learned from these experiences which may apply to future Moon bases.

Here is a map of all of the Moon landings by unmanned probes and manned landers through late 2019:

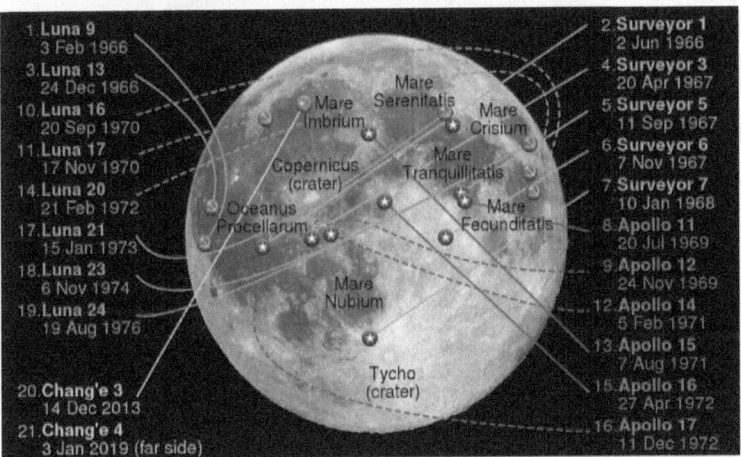

One thing the Moon landings do certainly provide is the confidence that if we got there once we can certainly go there again. Also, with the advancement of technology in fifty years since the first Moon landing we can do a lot more on the Moon than used to be possible.

One of the great limitations of the original manned lunar landings was that they usually had to be within five degrees of the lunar equator. The reason was the limited launch capabilities of one Saturn Five rocket. I know that the Saturn Five is the largest and most powerful rocket built so far, but having more landing flexibility means launching more rockets to carry more materials and fuel to the Moon.

Previous to manned Moon landings some scientists also thought that the Moon would be covered with a sea of Moon dust. This turned out not to be true and a lunar rover was used on later manned flights to travel around the surface.

Since the Apollo landings there have been numerous additional missions to the Moon by unmanned space probes to determine gravity maps, electromagnetic fields, and where water ice is located. Water Ice was mapped to be at both the North and South Poles. (See the picture below)

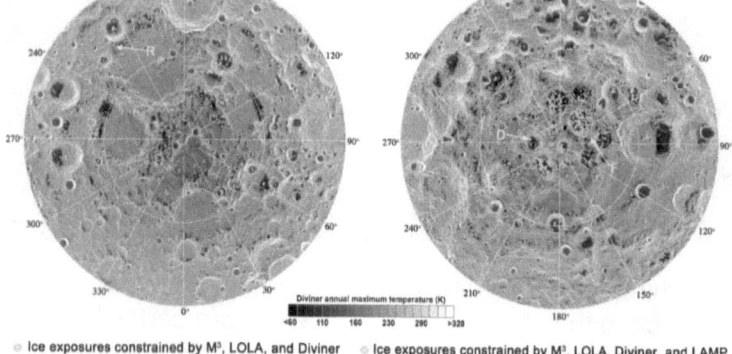

Ice exposures constrained by M³, LOLA, and Diviner   Ice exposures constrained by M³, LOLA, Diviner, and LAMP

There are probably also new surprises awaiting us since a Chinese mission to the far side just found some material which acts like a "Gel".

Gel on the Far side of the Moon

Last July, (2019) China's Yutu-2 rover discovered something with an unexpected color and luster during its travels on the far side of the Moon. On September 1, a tweet from People's Daily – largest newspaper group in China – used the words "gel-like" to describe this substance. Weird! The choice of words piqued a lot of curiosity, although some scientists stated at the time the rover had probably stumbled on something more like impact glass, created after a meteorite hits the lunar surface.

Now, it appears those scientists were right. The China Lunar Exploration Program has released a new photo of the substance, and the bright specks do resemble other impact glass – known as impactite and resembling trinitite on Earth – that's been seen on the Moon before. The photo, taken by the Yutu-2 rover's main camera, shows the center of the small crater, with numerous small bright spots on the lunar regolith.

The image doesn't look too unusual, just showing the grey regolith with the small bright flecks in the center of the crater. It was analyzed and processed to bring out more detail by Daniel Moriarty, a NASA Postdoctoral Program fellow at the Goddard Space Flight Center. As he explained:

> *The shape of the fragments appears fairly similar to other materials in the area. What this tells us is that this material has a similar history as the surrounding material. It was broken up and fractured by impacts on the lunar surface, just like the surrounding soil.*
>
> *I think the most reliable information here is that the material is relatively dark. It appears to have brighter material embedded within the larger, darker regions, although there is a chance that is light glinting off a smooth surface. But we're definitely looking at a rock.*

7.0 Recent Approaches for Moon Travel & Bases

The current approach about going to the Moon is to provide an infrastructure to not only land on the Moon but have a transfer station in orbit and build a base on the Moon. Under current directives a Moon base is seen as a learning experience for an eventual manned mission to Mars.

Here is the progression of ideas on the current approach to the Moon supported by NASA as of winter 2019:

7.1 The Deep Space Gateway

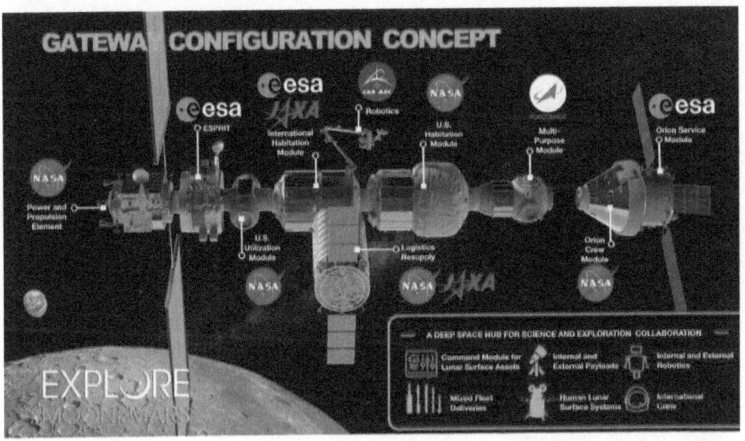

The Gateway will be an international effort and as of fall 2019 over twenty five countries want to sign up to participate.

The Lunar Gateway is designed to provide a way station for trips to land on the Moon. It will be in an elongated orbit around the Moon positioned such that landing craft can leave the gateway to land on different parts of the Moon from different orbital positions of the Gateway.

Lunar Orbit Details:

The Lunar Gateway is planned to be deployed in a highly elliptical seven-day near-rectilinear halo orbit (NRHO) around the Moon, which would bring the station within 3,000 km (1,900 mi) of the lunar north pole at closest approach and as far away as 70,000 km (43,000 mi) over the lunar south pole. Traveling to and from cislunar space (lunar orbit) is intended to develop the knowledge and experience necessary to venture beyond the Moon and into deep space.

The proposed NRHO orbit would allow lunar expeditions from the Gateway to reach a low polar orbit with a delta-v of 730 m/s and a half a day of transit time. Orbital station-keeping would require less than 10 m/s of delta-v per year, and the orbital inclination could be shifted with a relatively small delta-v expenditure, allowing access to most of the lunar surface. Spacecraft launched from Earth would perform a powered flyby of the Moon (delta-v = ~180 m/s) followed by a ~240 m/s delta-V NRHO orbit insertion burn to dock with the Gateway as it approaches the apoapsis point of its orbit. The total travel time would be 5 days; the return to Earth would be similar in terms of trip duration and delta-V requirement if the spacecraft spends 11 days at the Gateway. The crewed mission duration of 21 days and ~840 m/s delta-V are limited by the capabilities of the Orion life support and propulsion systems.

Lunar Gateway Modules are planned as follows:

<u>Contracted modules</u>

A) The **Power and Propulsion Element (PPE)** started development at the Jet Propulsion Laboratory during the now canceled Asteroid Redirect Mission. The original concept was a robotic, high performance solar electric spacecraft that would retrieve a multi-ton boulder from an asteroid and bring it to lunar orbit for study. When ARM was cancelled, the solar electric propulsion was repurposed for the Gateway. The PPE will allow access to the entire lunar surface and act as a space tug for visiting craft. It will also serve as the command and communications center of the Gateway. The PPE is intended to have a mass of 8-9 tons and the capability to generate 50 kW of solar electric power for its ion thrusters, which can be supplemented by chemical propulsion. It is currently planned to launch on a commercial launch vehicle in 2022. In May 2019, Maxar Technologies was contracted by NASA to manufacture this module, which will also supply the station with electrical power and is based on Maxar's 1300 series satellite bus. The PPE will use Advanced Electric Propulsion System (AEPS) Hall-effect thrusters. Maxar was awarded a firm-fixed price contract of $375 million to build the PPE. NASA is supplying the PPE with an S-band communications system to provide a radio link with nearby vehicles and a passive docking adapter to receive the Gateway's future utilization module.

B) The **Habitation and Logistics Outpost (HALO),** also called the Minimal Habitation Module (MHM) and formerly known as the Utilization Module, will be built by Northrop Grumman Innovation Systems (NGIS). A commercial launch vehicle would launch the HALO before the end of year 2023. The HALO is based on a Cygnus Cargo resupply module to the outside of which radial docking ports, body mounted radiators (BMRs), batteries and communications antennae will be added. The HALO will be a scaled-down habitation module, yet, it will feature a functional pressurized volume providing sufficient command, control & data handling capabilities, energy storage and power distribution, thermal control, communications and tracking capabilities, two axial and up to two radial docking ports, stowage volume, environmental control and life support systems to augment the Orion spacecraft and support a crew of four for at least 30 days.

C) The **European System Providing Refueling, Infrastructure and Telecommunications (ESPRIT) service module** will provide additional xenon and hydrazine capacity, additional communications equipment, and an airlock for science packages. It will have a mass of approximately 4 tons (8,800 lb), and a length of 3.91 m (12.8 ft). The studies and design are being performed mostly by Airbus and OHB. The module construction was approved in November 2019.

D) The **International Habitation Module (iHAB)** will be an additional habitation module built by ESA in collaboration with Japan. Together with the HALO module, they will provide a combined 125 m3 (4,400 cu ft) of habitable volume to the station.

Proposed modules

The concept for the Lunar Gateway is still evolving, and these modules have also been proposed to be added to the design:

The **Gateway Logistics Modules** will be used to refuel, resupply and provide logistics on board the space station. The first logistics module sent to the Gateway will also arrive with a robotic arm, which will be built by the Canadian Space Agency.

The **Gateway Airlock Module** will be used for performing extravehicular activities outside the space station and would have the docking port for the proposed Deep Space Transport.

All About Moon Bases-And Our Plans to Return to the Moon

7.2 Lunar Lander Concepts

Lunar landers for the next phase of Moon landings will not be one time landers but reusable vehicles.

In October 2020, NASA officials wants to select two contractors from the design study award winners. Those firms will proceed with full development of their human-rated lunar landers, and NASA will later choose one for a landing attempt in 2024, and another for a Moon landing in 2025.

NASA officials want to conduct lunar landing missions on a cadence of at least one per year after 2024.

Two astronauts are expected to be aboard for the Artemis program's first two landing attempts, including — as NASA regularly mentions — the first woman to land on the Moon.

Among other requirements, the landers for the 2024 and 2025 missions must provide the following capabilities:

- At least 1,907 pounds (865 kilograms) of payload delivered to the lunar surface, with a goal of 2,127 pounds (965 kilograms)
- At least 6.5 days on the lunar surface
- At least two spacewalks per mission, with goal of five spacewalks, using NASA-provided spacesuits
- At least 77 pounds (35 kilograms) of sample return capability, with a goal of 220 pounds (100 kilograms)

NASA will pick one or both of the lunar lander developers to work on a more "sustainable" lander design, a craft that will be able to carry at least four astronauts to the Moon's surface, operate during the two-week-long lunar night, and support longer-duration spacewalks. The more advanced lander could be ready to fly to the Moon in 2026, and must utilize NASA's Gateway in lunar orbit.

7.3 Settling the South Pole

Most current proposals for settling the Moon are focused on Shakelton Crater at the Lunar South Pole. Why? Because previous sensing satellites have found a large amount of Water ice at this location. The hope is that water ice can be minded and refined to provide water for habitation and decomposed for rocket fuel.

7.4 Far Side Bases

There are some good reasons to build bases on the far side of the Moon which never faces the Earth. One is radio astronomy which would be protected from all of Earth's electromagnetic emissions. Another reason are large telescopes. Since the Moon doesn't have any atmosphere they can be built on the far side which is both very dark and with no atmospheric distortion of telescopic images. These images would be incredible and since it will be on the Moon it can be serviced and improved as needed.

8.0 Types of Moon Bases

There are different ideas for building a Moon Base which all have positive and negative benefits and detractions.

Here are some of the major concepts below:

8.1 Using Lava Tubes

Lunar lava tubes are ancient volcanic tunnels on the Moon that are thought to have formed during basaltic lava flows. When the surface of a lava tube cools, it forms a hardened lid that contains the ongoing lava flow beneath the surface in a conduit-shaped passage. Once the flow of lava diminishes, the tunnel may drain, forming a hollow void.

Lunar lava tubes are formed on surfaces that have a slope that ranges in angle from 0.4° to 6.5°. Lunar lava tubes may be as wide as 500 metres (1,600 ft) before they become unstable against gravitational collapse. However, stable tubes may still be disrupted by seismic events or meteoroid bombardment.

The existence of a lava tube is sometimes revealed by the presence of a "skylight", a place in which the roof of the tube has collapsed, leaving a circular hole that can be observed by lunar orbiters. Here are some pictures of possible lava tubes on the Moon:

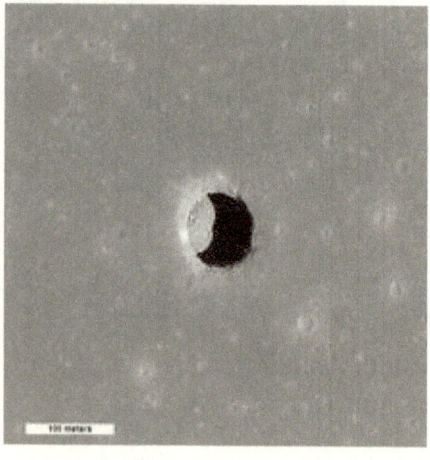

The next picture shows a possible tunnel on the Moon:

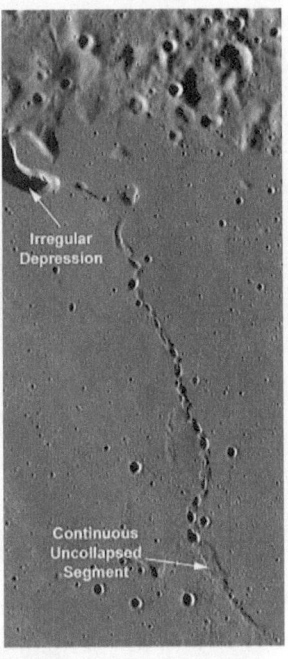

Using manmade tunnels and natural lava tubes for a structure has the benefits of good protection from solar radiation and making it easier to trap an atmosphere by just enclosing two sections of the tunnel. The problems have to do with the equipment needed for any tunnel digging needed and having to find a lava tube to build a base which might be in an inconvenient location.

8.2 Printing Buildings

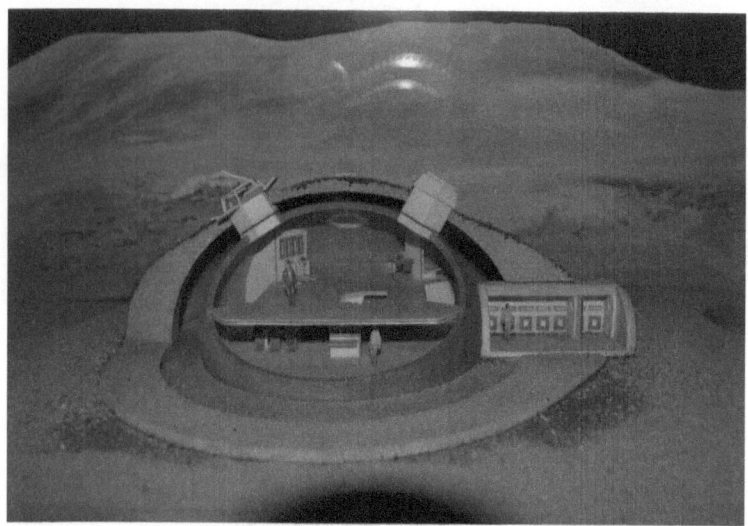

Three dimensional printers were developed from the concept of ink jet paper printers. What if you could use other materials than ink to deposit and build up an object layer by layer from a computer design?

Three D printers are now used widely around the world and are being used to print even complex objects like rocket engines.

Some people have even experimented with using this approach to build homes or other buildings on Earth.

This had led some Moon Colony designers to think about just using Lunar Regolith and a printing robot to build an entire shelter or pile up and compress regolith over a balloon structure. Maybe a combination of the two concepts will lead to the least expensive and quick to build lunar base.

8.3 Underground Structures

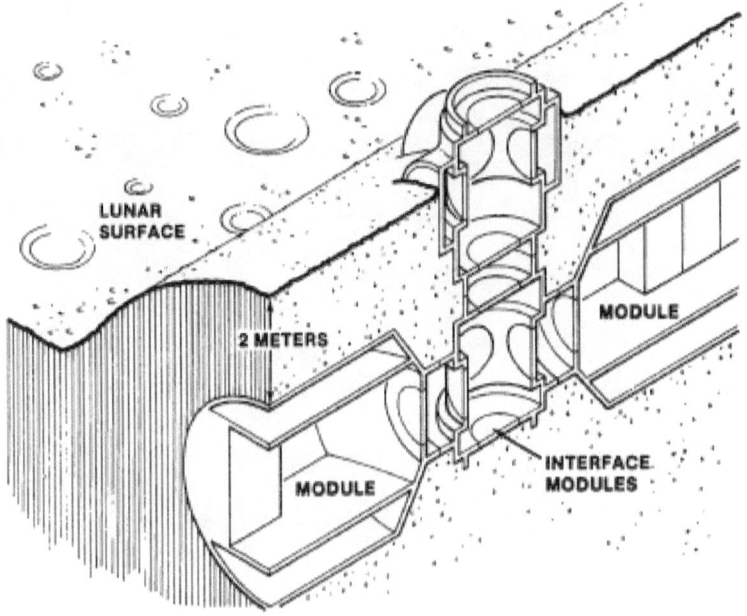

Building a structure underground has the benefits of radiation protection and makes it easier to contain an atmosphere. The problem is that it takes a lot of digging equipment to either dig pits for module structures or to dig tunnels for those structures. This is a more expensive approach to building an early Moon base than building on the surface. However, underground structures do have the benefit of being safe from radiation and easier to control atmospheric losses.

All About Moon Bases-And Our Plans to Return to the Moon

9.0 Other Needed Lunar Equipment

<u>Lunar Space Suits</u>

New space suits are being designed since the Apollo Moon suit design is over fifty years old.

The two spacesuit prototypes which NASA showcased are designed for two separate parts of a crewed mission to the Moon. One, called the Exploration Extravehicular Mobility Unit (xEMU) is a red, white and blue suit designed to be worn by astronauts exploring the lunar surface, specifically at the Moon's South Pole — the target for NASA's next crewed lunar landing.

The second suit is the Orion Crew Survival System, which is a bright orange pressure suit that will be worn by astronauts when they launch into space on the Orion capsule and return to Earth.

Vehicles

Plenty of automakers make off-road vehicles. Toyota is looking to make one for off-world use.

The carmaker is teaming up with Japan's national space agency to develop and build a Moon rover that future astronauts will use to explore the lunar surface.

Unlike the trio of bare-bones NASA rovers that Apollo astronauts steered across the lunar surface in the 1970s, the proposed rover will have an enclosed, pressurized cabin. In artist renderings, the vehicle looks a bit like a big off-world SUV.

"Manned, pressurized rovers will be an important element supporting human lunar exploration, which we envision will take place in the 2030s," Koichi Wakata, vice-president of the Japan Aerospace Exploration Agency (JAXA), said in a written statement. He said the space agency aims to launch the new rover in 2029.

The rover will be about 20 feet long and 17 feet wide, and will feature a 140-cubic-foot cabin capable of accommodating two passengers — four in an emergency. It will be powered by fuel cell technology similar to what's used in some of Toyota's earthbound vehicles. Fuel cells run on oxygen and hydrogen and emit only water.

The rover will have a range of more than 6,200 miles, according to Toyota. That represents a big improvement over the NASA lunar buggies brought to the Moon on the Apollo 15, 16 and 17 missions. Those electric vehicles were designed only for short trips; Apollo 17 astronauts Gene Cernan and Jack Schmidt set the distance record, driving their rover a total of 22.3 miles on three separate outings in December 1972.

<u>Power Production</u>

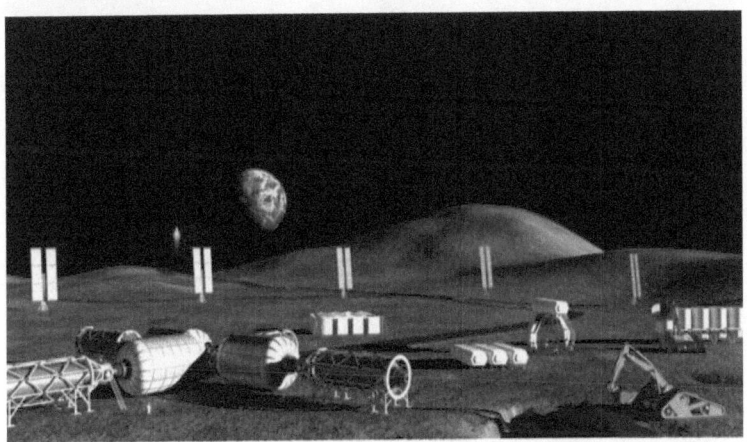

The preferred system recommended in the 2009 NASA study was a photovoltaic solar array-powered cryogenic storage regenerating fuel cell system. NASA calculated that a five-kilowatt continuous delivery system would store 2,000 kilowatt-hours with a system energy density of 1.15 kilowatt-hours per kilogram. The study's alternate preferred system was a fixed orbit laser system, with a 16.1-hour orbit period that required a surface receiver installation with 525 kilowatt-hours of energy storage. The laser was powered and fired when it was both in direct sunlight and in direct line-of-sight with the Moon base.

Communications

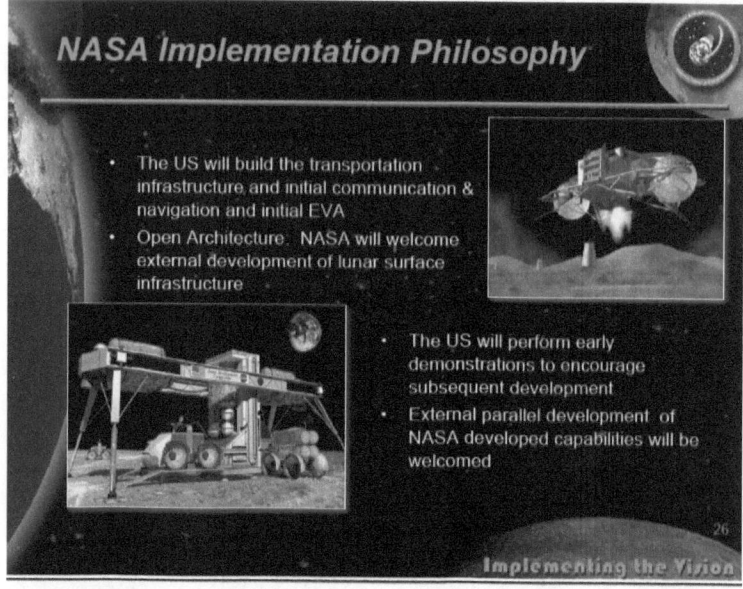

Studies done by NASA in 2007 suggest that there be a lunar orbiting communication or satellite or multiple satellites to provide a relay for communications with the Lunar Gateway and Earth. This would also be a digital network to provide IP internet messaging capabilities. Updated studies still need to be done but this one provides a good first approach to providing communications.

Mining Water Ice

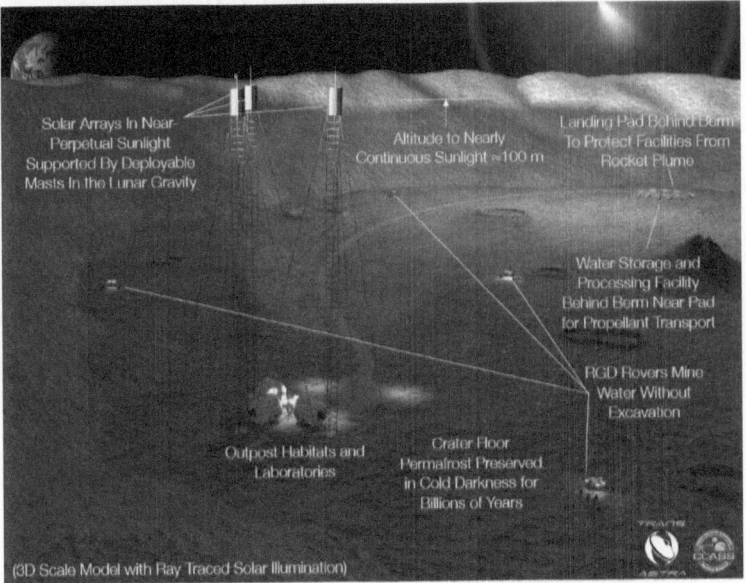

The Lunar Polar Gas-Dynamic Mining Outpost (LGMO) is a breakthrough mission architecture that promises to greatly reduce the cost of human exploration and industrialization of the Moon. LGMO is based on two new innovations that together solve the problem of affordable lunar polar ice mining for propellant production.

The first innovation is based on a new insight into lunar topography: the analysis suggests that there are large (hundreds of meters) landing areas in small (0.5-1.5 km) near polar craters on which the surface is permafrost in perpetual darkness but with perpetual sunlight available at altitudes of only 10s to 100s of meters. In these prospective landing sites, deployable solar arrays held vertically on masts 100 meters or so in length (lightweight and feasible in lunar gravity) can provide nearly continuous power. This means that a large lander, such as the Blue Moon vehicle proposed by Blue Origin, a BFR; or a modestly sized lunar ice mining outpost could sit on mineable permafrost with solar arrays in perpetual sunlight on masts providing affordable electric power without the need to separate power supply from the load.

The second enabling innovation for LGMO is Radiant Gas Dynamic (RGD) mining. RGD mining is a new Patent Pending technology invented by TransAstra to solve the problem of economically and reliably prospecting and extracting large quantities (1,000s of tons per year) of volatile materials from lunar regolith using landed packages of just a few tons each. To obviate the problems of mechanical digging and excavation, RGD mining uses a combination of radio frequency, microwave, and infrared radiation to heat permafrost and other types of ice deposits with a depth-controlled heating profile. This sublimates the ice and encourages a significant fraction of the volatiles to migrate upward out of the regolith into cryotraps where it can be stored in liquid form. RGD mining technology is integrated into long duration electric powered rovers.

In use, the vehicles stop at mining locations and lower their collection domes to gather available water from an area before moving on. When on-board storage tanks are full, the vehicles return to base to empty tanks before moving back out into the field to continue harvesting. The rover can be battery operated and recharge at base or carry a laser receiver powered by a remote laser. Based on these innovations, LGMO promises to vastly reduce the cost of establishing and maintaining a sizable lunar polar outpost that can serve first as a field station for NASA astronauts exploring the Moon, and then as the beachhead for American lunar industrialization, starting with fulfilling commercial plans for a lunar hotel for tourists.

Tools and Equipment

The tools necessary to carry out any significant production of raw materials for manufacturing cannot be regarded as simple, but merely as relatively simple.

One item will have to be dirt-moving equipment, necessary for habitat excavation, as well as for transporting feedstocks to smelting or manufacturing sites and removing waste products and, perhaps, the products themselves. It is assumed that all processes, from gathering and production of raw materials through manufacture and assembly, will be substantially automated. Equipment probably will be tended, mainly from remote stations nearby.

Equipment should be designed to allow straightforward repair, optimization to actual encountered conditions, and innovative adaptation by the operator to new feedstocks and conditions. Electrical power, at least one megawatt, will have to be available for any significant processing or manufacturing activity. Initially, this probably will need to be nuclear power, although eventually solar power should be exploited to the fullest extent. Through the use of concentrating mirrors, solar power should be available at the outset for heating of materials to high enough temperatures to melt or even distill them.

All About Moon Bases-And Our Plans to Return to the Moon

10.0 Current Plans for Moon Bases

Here is the core of a recent article from Spring 2019 on NASA's plans for a Moon base:

NASA's official plans to build a permanent base on the Moon have leaked online, revealing how and when astronauts will return to the rocky world for the first time in 50 years.

Internal documents show how NASA wants to launch 37 rockets to the Moon within the next decade, with at least five of these carrying astronauts.

Starting with an unmanned rover in 2023, the space agency is expected to land people on the Moon in 2024.

NASA will then fire manned missions to Earth's neighbor every year between 2024 and 2028, according to the documents, which were obtained by Arstechnica.

Speaking to The Sun, a NASA spokeswoman confirmed the documents are real and revealed the plans were briefed today during a public session of the Science Committee to the Nasa Advisory Council (NAC).

They show a decade-long program that culminates with a permanent lunar base, which NASA will begin building in 2028.

They are in part a response to recent calls from U.S. Vice President Mike Pence to take astronauts back to the Moon.

"In the nearly two months since Pence directed Nasa to return to the Moon by 2024, space agency engineers have been working to put together a plan that leverages existing technology, large projects nearing completion, and commercial rockets to bring this about," Arstechnica's Eric Berger wrote.

"Last week, an updated plan that demonstrated a human landing in 2024, annual sorties to the lunar surface thereafter, and the beginning of a Moon base by 2028, began circulating within the agency."

Berger did not say how he obtained the plans, which have not yet been made public.

They do appear to line up with previous statements from NASA about its lunar program, codenamed Artemis.

As with any space exploration project, the main obstacle is cash.

NASA reckons it will need $4.7 billion to $8.2 billion per year on top of NASA's existing budget of about $20 billion.

Boss Jim Bridenstine recently asked for an extra $1.6 billion in fiscal year 2020 to start developing a lunar lander.

The plan also relies heavily on contractors delivering ambitious hardware on time, which has hindered NASA in the past.

Boeing has been developing the core stage of the agency's next-gen rocket, the Space Launch System, for eight years – but has yet to come up with the goods.

All About Moon Bases-And Our Plans to Return to the Moon

11.0 Technologies to Use at the Moon Base

Many technologies which were already developed and prototyped on the International Space Station can also be used on a Moon Base:

Heat and Air Conditioning

The ISS has a lot of design elements used to maintain and control temperature. Without thermal controls, the temperature of the orbiting Space Station's Sun-facing side would soar to 250 degrees F (121 C), while thermometers on the dark side would plunge to minus 250 degrees F (-157 C). There might be a comfortable spot somewhere in the middle of the Station, but searching for it wouldn't be much fun!

Fortunately for the crew and all the Station's hardware, the ISS is designed and built with thermal balance in mind -- and it is equipped with a thermal control system that keeps the astronauts in their orbiting home cool and comfortable. The first design consideration for thermal control is insulation -- to keep heat in for warmth and to keep it out for cooling.

Here on Earth, environmental heat is transferred in the air primarily by conduction (collisions between individual air molecules) and convection (the circulation or bulk motion of air). "This is why you can insulate your house basically using the air trapped inside your insulation," said Andrew Hong, an engineer and thermal control specialist at NASA's Johnson Space Center. "Air is a poor conductor of heat, and the fibers of home insulation that hold the air still minimize convection."

"In space there is no air for conduction or convection," he added. Space is a radiation-dominated environment. Objects heat up by absorbing sunlight and they cool off by emitting infrared energy, a form of radiation which is invisible to the human eye.

As a result, insulation for the International Space Station doesn't look like the fluffy mat of pink fibers you often find in Earth homes. The Station's insulation is instead a highly-reflective blanket called Multi-Layer Insulation (or MLI) made of Mylar and Dacron.

The reflective silver mesh is aluminized Mylar. The copper-colored material is kapton, a heavier layer that protects the sheets of fragile Mylar, which are usually only 0.3 mil or 3/10000 of an inch thick.

"The Mylar is aluminized so that solar thermal radiation can't get through it," explains Hong. Here on Earth, we use blankets containing aluminized Mylar to wrap people who have been exposed to cold or trauma. Such blankets are especially popular among hunters and campers!

"Layers of Dacron fabric keep the Mylar sheets separated, which prevents heat from being conducted between layers," he continued. "This ensures radiation will be the most dominant heat transfer method through the blanket."

Except for its windows, most of the ISS is covered with the radiation-stopping MLI.

Water Purification

All water used to be hauled into space by rocket then used up or wasted. The ISS now has a water production system in usage since 2010.

Drinkable water is one of the primary and most important assets for human survival. So when preparing for a journey, whether to sea or to space, planners must take this vital resource into consideration. Stowage space during such voyages always comes at a premium. It is no different for the International Space Station and the resupply vehicles that dock there.

A great example of a solution to minimize size and weight in life support is the recently launched Sabatier system. Originally developed by Nobel Prize-winning French chemist Paul Sabatier in the early 1900s, this process uses a catalyst that reacts with carbon dioxide and hydrogen - both byproducts of current life-support systems onboard the space station - to produce water and methane. This interaction closes the loop in the oxygen and water regeneration cycle. In other words, it provides a way to produce water without the need to transport it from Earth.

The fundamental technology for this particular system has been in development for the past twenty years. The overall schedule for hardware production, however, was under two years. This accelerated timeline was a significant challenge for the complex Sabatier, which contains a furnace, a multistage compressor, and a condenser/phase-separation system. The fact that recycling system feeds for Sabatier were already available on the station helped to simplify some of the design tasks by reducing the unknowns.

According to Jason Crusan, chief technologist for space operations at NASA Headquarters in Washington, the previous development and solid interfaces allowed NASA to try out a new way of acquiring services for the station with Sabatier. "Being able to demonstrate innovative new methods to acquire technical capabilities is one of the key cornerstones the space station can serve for future missions and approaches to those missions," Crusan explained.

Using developing technologies and productive systems enables the station to squeeze every drop from the resources that must launch from Earth. In addition to improving the efficiency of the station's resupply capabilities, Sabatier also frees up storage space. This helps to maximize the area available for science facilities and engineering equipment. The knowledge gained from such systems also advances the collective understanding of technologies to advance spaceflight and help solve similar problems on Earth.

The Sabatier system has long been a part of the space station plan, but the retirement of NASA's space shuttles elevated the need for new resources to provide water. For a decade, shuttles have provided water for the station as a byproduct of the fuel cells they use to generate electricity. Sabatier supplements the capability of resupply vehicles to provide water to the station, without becoming a sole source for this critical station resource.

Currently in operation on the station, Sabatier is the final piece of the regenerative environmental control and life-support system. This hardware was successfully activated in October 2010 and interacts directly with the Oxygen Generation System, which provides hydrogen, sharing a vent line.

Prior to Sabatier, the Oxygen Generation System vented excess carbon dioxide and hydrogen overboard. Rather than wasting these valuable chemicals, Sabatier enables their reuse to generate additional water for the station. With room and resources at a premium in space, this is a significant contribution to the space station's supply chain.

In addition there is now a degree of water recycling on the ISS. Nature's been recycling water on Earth for eons, and now NASA is set to do the same thing above Earth on the International Space Station. Space shuttle Endeavour carried in two refrigerator-sized racks packed with a distiller and an assortment of filters designed to process astronauts' urine and sweat into clean drinking water.

The station crew depends now on water carried up aboard a space shuttle or cargo rocket. But an operational water recycler is expected to cut that need by 65 percent by producing about 6,000 pounds of potable water each year. That's enough fresh water to allow the station to host six crew members instead of three.

A system that operates on the station also will provide a significant stepping stone to developing even more efficient processes that will support astronauts on the moon or on long-duration voyages into the solar system. Although Russia's space station Mir recycled cosmonaut's sweat, the NASA recycler is the first to be flown in space that intends to cleanse and reuse almost all the water a crew member produces.

The system can recycle about 93 percent of the water it receives, said Bob Bagdigian, the Environmental Control Life Support System project manager at NASA's Marshall Space Flight Center in Huntsville, Ala. The water recycler counts in large part on a distiller that Bagdigian compares to a keg tilted on its side. On Earth, distilling is a simple process of simply boiling water and cooling the steam back into pure water. But without gravity, the contaminants in water never separate from the steam no matter how much heat is used.

"In space, it becomes quite a challenge to distill any liquid in the absence of gravity," Bagdigian said.

So the keg-sized distiller is spun up to produce an artificial gravity field. The contaminants in the urine press against the sides of the drum while the steam gathers in the middle and is pumped to a filter. The filters are not much different from those used on Earth, which means they use charcoal-like materials to pull more unwanted elements from the water. Another process uses chemical compounds that bond with the remaining contaminants so filters can pick them out of the water, too.

"The water that we produce meets or exceeds most municipal water product standards," Bagdigian said. The system has been in different stages of development ever since NASA committed to building a space station in the 1980s. Along the way, individual parts of the system have been flown on space shuttle missions for tests.

The distiller mechanism flew in 2003 and worked just fine in orbit, Bagdigian said. Now the crew of the International

Space Station will test the whole apparatus, but they won't drink any at first. Instead, they will take numerous samples and return them to Earth for detailed testing. After the testing is complete, controllers will clear the astronauts to use the fresh water in orbit.

NASA's water filter development has also helped produce filters that are now used in humanitarian efforts to make clean water in areas served only by contaminated sources. The effort to make a crew support system that reduces the need for fresh supplies from Earth includes an oxygen generator that is already installed in NASA's Destiny lab on the space station.

Housed in one rack instead of the two required for the water recycler, the oxygen producer splits the oxygen and hydrogen molecules in water and sends the oxygen into the space station as breathable air. The hydrogen is now dumped overboard. However, another process is under development that will combine the hydrogen with other chemicals that react with each other and produce more water.

While the water recycler in use will work fine for the International Space Station's needs, Bagdigian said work is already under way to make it more efficient so it can be used on long moon exploration missions. "We'll take this system and continue to push its performance and efficiency," Bagdigian said.

Solar Array Wings

(The below systems can be adapted from ISS usage experiences)

The electrical system of the International Space Station is a critical resource for the ISS because it allows the crew to live comfortably, to safely operate the station, and to perform scientific experiments. The ISS electrical system uses solar cells to directly convert sunlight to electricity. Large numbers of cells are assembled in arrays to produce high power levels. This method of harnessing solar power is called photovoltaics.

The process of collecting sunlight, converting it to electricity, and managing and distributing this electricity builds up excess heat that can damage spacecraft equipment. This heat must be eliminated for reliable operation of the space station in orbit. The ISS power system uses radiators to dissipate the heat away from the spacecraft. The radiators are shaded from sunlight and aligned toward the cold void of deep space.

Each ISS solar array wing (often abbreviated "SAW") consists of two retractable "blankets" of solar cells with a mast between them. Each wing uses nearly 33,000 solar cells and when fully extended is 35 meters (115 ft.) in length and 12 meters (39 ft.) wide. When retracted, each wing folds into a solar array blanket box just 51 centimeters (20 in) high and 4.57 meters (15.0 ft.) in length. The ISS now has the full complement of eight solar array wings. Altogether, the arrays can generate 84 to 120 kilowatts.

The solar arrays normally track the Sun, with the "alpha gimbal" used as the primary rotation to follow the Sun as the space station moves around the Earth, and the "beta gimbal" used to adjust for the angle of the space station's orbit to the ecliptic. Several different tracking modes are used in operations, ranging from full Sun-tracking, to the drag-reduction mode ("Night glider" and "Sun slicer" modes), to a drag-maximization mode used to lower the altitude.

Batteries

Since the station is often not in direct sunlight, it relies on rechargeable nickel-hydrogen batteries to provide continuous power during the "eclipse" part of the orbit (35 minutes of every 90 minute orbit). The batteries ensure that the station is never without power to sustain life-support systems and experiments. During the sunlit part of the orbit, the batteries are recharged. The nickel-hydrogen batteries have a design life of 6.5 years which means that they must be replaced multiple times during the expected 20-year life of the station. The batteries and the battery charge/discharge units are manufactured by Space Systems/Loral (SS/L), under contract to Boeing. N-H2 batteries on the P6 truss were replaced in 2009 and 2010 with more N-H2 batteries brought by Space Shuttle missions. There are batteries in Trusses P6, S6, P4, and S4.

Since 2017, nickel-hydrogen batteries are being replaced by lithium-ion batteries. On January 6, a multi-hour EVA began the process of converting some of the oldest batteries on the ISS to the new lithium-ion batteries There are a number of differences between the two battery technologies, and one difference is that the lithium-ion batteries can handle twice the charge, so only half as many lithium-ion batteries are needed during replacement. Also, the lithium-ion batteries are smaller than the older nickel-hydrogen batteries. Although they are not quite as long lasting as nickel-hydrogen, they can last long enough to extend the life of ISS.

ISS Electrical Power Distribution

The power management and distribution subsystem operates at a primary bus voltage set to Vmp, the peak power point of the solar arrays. As of 30 December 2005, Vmp was 160 volts DC (direct current). It can change over time as the arrays degrade from ionizing radiation. Microprocessor-controlled switches control the distribution of primary power throughout the station.

The battery charge/discharge units (BCDUs) regulate the amount of charge put into the battery. Each BCDU can regulate discharge current from two battery ORUs (Orbital Replacement Unit, a series-connected pack of 38 Ni-H2 cells), and can provide up to 6.6 kW to the Space Station. During insolation, the BCDU provides charge current to the batteries and controls the amount of battery overcharge. Each day, the BCDU and batteries undergo sixteen charge/discharge cycles. The Space Station has 24 BCDUs, each weighing 100 kg.

Data Architecture/Communications

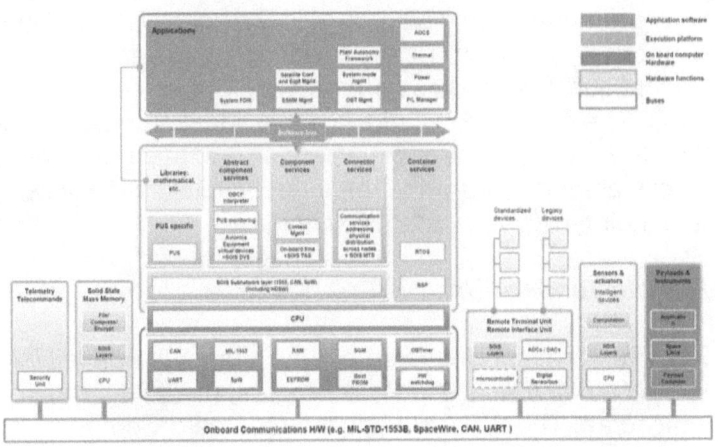

The ISS data architecture and communications system is very complex. I included an architecture diagram above and detailed overview below so that you can see just how much is involved.

(Similar systems can be used at the Moon Base.)

On a larger habitat in space imagine that the architecture is that much more complex according to its size and the number of people on it. Fortunately, computing architecture is one area where continuous advances should keep up with the computing needs of an advanced space habitat facility.

Spacecraft Management Unit

On the ISS The On-board Computer (also referred to as Spacecraft Management Unit - SMU or Command & Data Handling Management unit - CDMU) is the central core of the Spacecraft Avionics. The Central Processing Unit

(CPU) hosts the Execution Platform software (composed of RTOS, BSP, SOIS layers, PUS, …) and the Application software. Volatile and Non-volatile Memories, Safe Guard Memories, On Board Timer, Interface controllers and Reconfiguration modules are the other main blocks of a OBC. The figure above shows a functional architecture of the On-Board Data System where all the major functional blocks are indicated with their intercommunication links and their typical redundancy scheme.

Remote Terminal Unit

Remote Terminal Unit (also called Remote Interface Unit-RIU) is a unit that is usually present on medium-large size spacecraft. The RTU offloads the On Board Computer from analogue and discrete digital data acquisition and actuators control tasks.

Platform Solid State Mass Memory

For Earth Observation missions the mass memory for the P/L data may belong to the satellite platform and sometimes, depending on the capacity required, might be included inside the OBC as a single module.

TM/TC

The tele commands, once validated, are multiplexed to the intended addresses. There are two categories of commands: the high priority and the normal commands. The high priority commands (HPC) are sent to the Command Pulse Distribution Unit (CPDU) for immediate execution. The CPDU is either internal to the TC decoder or external and it's implemented in hardware, i.e. no software is involved in the execution of HPCs. The normal commands are sent off to the OBC CPU to be either processed or relayed on the system bus. The Telemetry encoder collects the Telemetry packets from different sources (processing, data storage, essential telemetry, payload), assembles the Telemetry transfer frames and sends them to the TM/TC transceiver to be downloaded to the ground.

Busses

The most common command and control bus used on a spacecraft platform is the MIL-STD-1553B covered by the ECSS-E-ST-50-13C. An alternative to the MIL-STD-1553B is the CAN that ESA and the European Space community is standardizing for space applications. UART serial channels are also used especially to control AOCS sensors. The Spacewire technology is now being increasingly used for data transfers < 160 Mbit/s and it can combine the command and control function with massive data transfer.

Communication protocols

The space community is asking for a real improvement in the specification and use of communications protocols. Typically, previous developments have harmonized physical interfaces and low level data link protocols but above this level proprietary solutions have been utilized. This has without any doubt increased development and integration costs and limited the possibility of element reuse without expensive modification. In comparison, the commercial market on the ground has systematically pursued the use of multilayer protocol stacks resulting in simple integration and multi-vendor compatibility. This commercial trend is now being adopted for the flight avionics by the development and standardization of protocols above the basic link layer.

All About Moon Bases-And Our Plans to Return to the Moon

12.0 Schedules for Construction

Schedules for a Lunar Base are still very preliminary and things will change over time. Here is one schedule document with speculations for future changes:

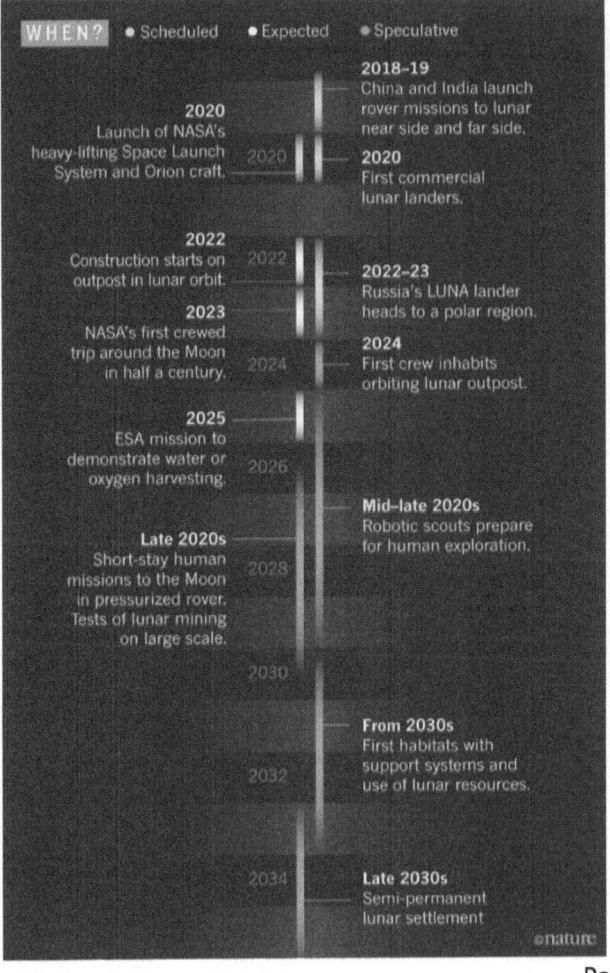

The first new Lunar Landing is scheduled for 2024 with a minimal orbiting sized Lunar Gateway to be used as a staging location for the lunar surface mission.

Successive visits to the Moon will then be scheduled yearly.

Lunar water harvesting will be tested in 2025

Short stays on the Moon will continue into the second half of the 2020s with first long term habitats to be built in the early 2030s.

13.0 Participating Countries and Companies

As of late 2019 over 25 countries have offered to participate in the Artemus Program as well as lots of major companies too. Here are some of the major players of both companies and countries:

<u>Space Launch System</u>

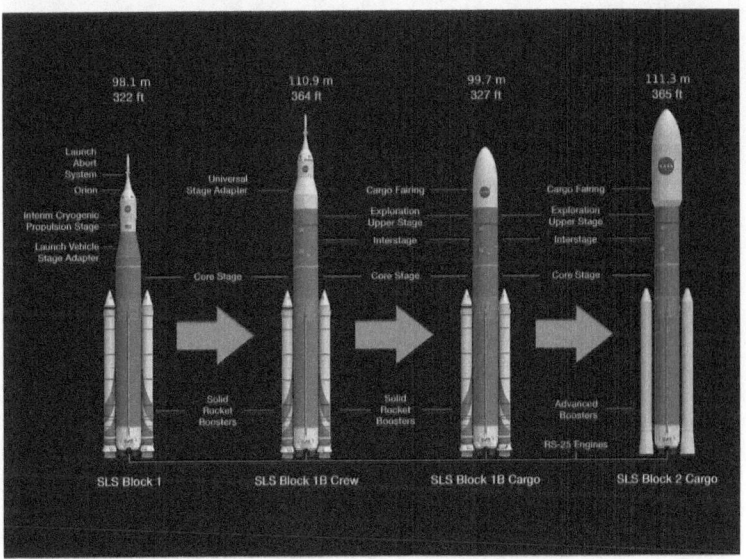

This system will launch the Lunar Gateway and Landers to the Moon

Boeing-They are the main contractor for the Space Launch System which is NASA's new heavy lift launch system to the Moon and beyond.

The Orion Capsule

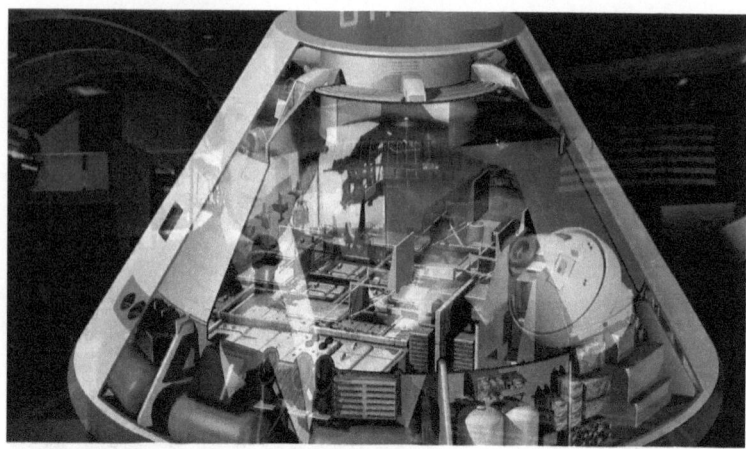

This new deep space capsule can carry a crew of two to six persons and is designed for Moon and deep space missions.

Lockheed-Martin-They are the prime contractor for the Orion Capsule.

The Orion Multi-Purpose Crew Vehicle (Orion MPCV) is a class of partially reusable spacecraft used in NASA's human spaceflight programs. Consisting two components – a Crew Module (CM) manufactured by Lockheed Martin, and a European Service Module (ESM) manufactured by Airbus Defence and Space – the spacecraft are designed to support crewed exploration beyond low Earth orbit. Orion is equipped with solar power, an automated docking system, and glass cockpit interfaces modeled after those used in the Boeing 787 Dreamliner, and can support a crew of six up to 21 days undocked and up to six months docked. A single AJ10 engine provides the spacecraft's primary propulsion, while eight R-4 D-11 engines and six pods of custom reaction control system engines developed by Airbus provide the spacecraft's secondary propulsion. Although compatible with other launch vehicles, Orion is primarily designed to launch atop a Space Launch System (SLS) rocket, with a tower launch escape system.

Orion was conceived by Lockheed Martin as a proposal for the Crew Exploration Vehicle (CEV) to be used in NASA's Constellation program. Lockheed Martin's proposal defeated a competing proposal by Northrop Grumman, and was selected by NASA in 2006 to be the CEV. Originally designed with a service module featuring a new "Orion Main Engine" and a pair of circular solar panels, the spacecraft was to be launched atop the Ares I rocket with the Max Launch Abort System equipped. Following the cancellation of the Constellation program in 2010, Orion was heavily redesigned for use in NASA's Journey to Mars initiative; later named Moon to Mars.

The SLS replaced the Ares I as Orion's primary launch vehicle, and the service module was replaced with a design based on the European Space Agency's Automated Transfer Vehicle. A development test article of Orion's CM was launched in 2014 during Exploration Flight Test-1. As of 2019, two Orion spacecraft are under construction, with an additional two ordered, for use in NASA's Artemis program – the first of these is due to be launched in 2020 during Artemis 1.

ESA/Airbus-Defense-They will provide the service module for the Orion capsule which will have large wings to generate solar power and provide other services to Orion.

The European Service Module (ESM) is the service module component of the Orion spacecraft, serving as its primary power and propulsion component until it is discarded at the end of each mission. In January 2013, NASA announced that the European Space Agency (ESA) will contribute the service module for Artemis 1, based on ESA's Automated Transfer Vehicle (ATV). After approval of the first module, further decisions will be made that the module be provided by ESA for missions Artemis 1 to Artemis 4 included.

The service module supports the crew module from launch through separation prior to reentry. It provides in-space propulsion capability for orbital transfer, attitude control, and high altitude ascent aborts. It provides the water and oxygen needed for a habitable environment, generates and stores electrical power, and maintains the temperature of the vehicle's systems and components. This module can also transport unpressurized cargo and scientific payloads.

Lunar Gateway Partners

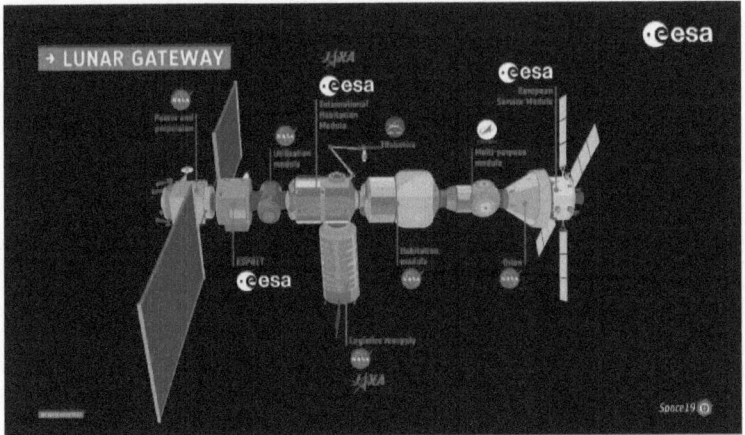

NASA-NASA will build a number of modules including:

- The Power and Propulsion Element
- The US Utilization Module
- The US Habitation Module

Jaxa-The Japanese Space Agency will be helping on the International Habitation Module and the Logistics Resupply Module

Canada-They will provide a robotic arm like they already have for the Space Shuttle and the International Space Station

ESA-The European Space Agency is building several key items. These include the Service Module for the Orion Space Capsule, the International Habitation Module for the Lunar Gateway, and the Esprit Module also for the Lunar Gateway.

Roscosmos-The Russian Space Agency will build the Lunar Gateway Multipurpose Module.

<u>Lunar Landers</u>

Blue Origin-Has offered one design for a Lunar Lander. This contract is still being bid and competed.

Lockheed-Martin-Has also proposed a Lunar Lander concept.

Additional Resource Providers

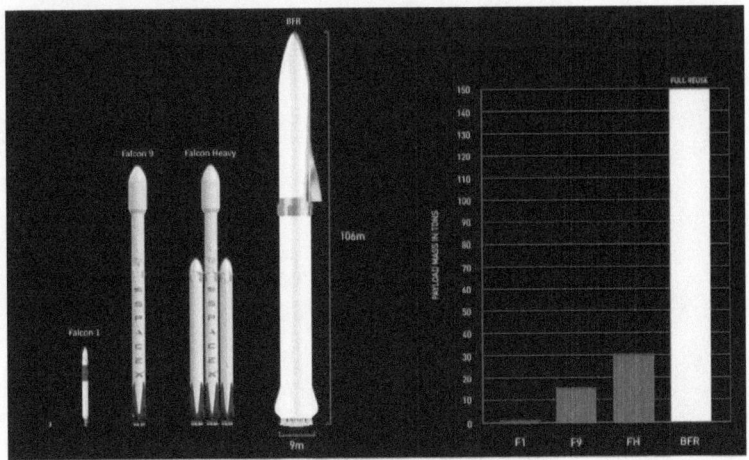

SpaceX-With its Falcon Heavy Rocket and planned BFR giant rocket plans to offer heavy lift capabilities for whatever additional transportation NASA needs for its Moon infrastructure and Moon Base.

14.0 Moon Base Cost Estimates

Estimated costs to ship materials to the Moon to start building the Moon Base will be very high. The Costs have been estimated in a very wide range.

Estimate A:

Some estimates done in 2007 say that reaching the Moon again would cost about $10 billion — estimates range from $7 billion and $13 billion — with an additional $28 billion to $52 billion being spent on the construction of base-related structures

Estimate B:

People have dreamt about living on the Moon for a long time, and while it is possible, it is also costly. Wendover Productions made a video explaining exactly how much it would cost to live on the Moon.

To calculate the amount, they figured out things such as how much a rocket would cost to use per pound, the cost of landers, the cost of a lunar base, and greenhouses for food. Their summary is that a grand total for four astronauts to live on the Moon for one year comes out to $36,000,000,000.

Estimate C:

In 2005, NASA estimated that returning humans to the Moon would cost $100 billion (approximately $122 billion in today's dollars). But if the success of private spaceflight companies like SpaceX and Orbital Sciences continues,

NASA could send humans back to the lunar surface in as little as five to seven years, at a highly reduced cost, the new report shows.

That's not all: 10 to 12 years after that first commercial Moon trip, NASA could develop a permanent base on the Moon for about $40 billion in today's dollars, the report said. The proposed permanent Moon base would be used to convert lunar ice into hydrogen propellant that could be sold for use by other spacecraft, including missions headed to Mars.

Launch costs included in the report were based on prices quoted for SpaceX's Falcon 9 and Falcon Heavy rockets.

The command/service module for astronauts was based on the human-rated Dragon spacecraft that SpaceX is developing for International Space Station missions.

<u>Cost Summary</u>

The estimates are all over the map but we can expect that building a Moon Base would have costs similar to building the International Space Station which was $100 billion dollars.

15.0 A Story of Building the Moon Base

My book "The Moon and Beyond" is still a work in progress but it includes a few chapters with the most realistic description I could provide of the what the process of building a Moon Base would be like. These chapters follow describing early Moon Base construction:

15.1 Early Moon Base Construction

We landed after several orbital corrections and a powered descent. The landing was anti-climactic with our engine kicking up lots of dust and then the motion stopped and we became aware that we were experiencing a one sixth Earth gravity. After making sure all the systems were operating properly, we opened the main hatch with us all wearing spacesuits to get outside.

The Commander and his assistant went down the ladder. Their first action was to activate the inflatable temporary structure at the base of the lander. This structure started to inflate and would provide us with shelter for the next several weeks as the main base was built. The inflatable dome was twenty feet in diameter and had an airlock built in. We each had a little sleeping cubicle and there was a galley and work areas. After it inflated we went inside to check airflow, heat, and then moved in supplies for living there.

We needed this shelter since living inside the lander for an extended period of time was a guarantee for crew stress and awful overcrowding.

We all took a walk around the landing site since we were all so excited to be there. The landing site was about one half mile from the crater wall and was in deep shade which was a couple of hundred degrees below zero Celsius. I could see the crater rim curving away to the horizon where it went out of sight. On the other side of the lander the land was just flat although we could see impact rocks in the distance. We could also see a glow over the crater rim because the Sun was shining on it from the other side. I was also thinking about the best location for the shelter. Did we want it out in the open or next to the crater's wall? Next to the crater wall would be more protected in the long run. However, I didn't need to think long because the building's site had already been picked out on Earth.

That night we had a little party in the temporary structure and all rested well before work was to start in the morning. Next morning Olga and I were the first ones outside. We needed to prepare the shelter site and setup the 3D construction equipment. There was a small tractor to be assembled which would be used to dig out a base for the building and make sure the foundation was firm. The tractor used a radio isotopic power source using Plutonium to create heat which was converted to electricity. The tractor had a radio control which we had practiced with back on Earth to control its movements.

I spent the next few hours digging a foundation pit with Olga relieving me as needed. After several days of effort we had a sufficient foundation dug and ready for construction. The foundation was round and one hundred feet in diameter. Our intent was to build a structure which could eventually hold fifty living and working people inside. The next step was the construction of the larger girders for the three dimensional construction machine. Before we could actually start building the building we needed girders to raise the machine above the ground and provide tracks to move the construction printing head over the construction. Imagine that we were building a large printer larger than the building size. The printing head would move in computer controlled movements over the ground to print the building underneath it. The preparation project took several days.

We had been on the Moon a week and we were dead tired at the end of each workday. The commander made sure we all ate dinner together in the temporary shelter to update each other and build a sense of community.

Captains Hold and Neemar had just gotten back from a field trip up to the crater rim. Their job was to install a communications antenna with repeaters aimed to our site back inside the crater. They used an open rocket powered vehicle to launch up to the rim and come back. At the top in the sunlight, and in line with Earth they installed solar panels and the antenna and communications equipment we needed to have regular communications with the DSG and Earth.

Stark and Springer were responsible for the ice mining. They had already done radar surveys within a couple miles radius of our landing site using a simple lunar rover and were now laying out the foundation for the mining and extraction site. Another mission would bring sufficient mining and purification equipment to start the generation of large quantities of water, and its extraction into hydrogen and oxygen.

15.2 Main Shelter Construction

Finally, after almost two weeks of foundation digging, pre-construction, and setup of the 3D printing equipment we were ready to get started.

The machine had a hopper where we would feed it with Moon materials of a granular type, and a silicate based binder which would bind it all together like concrete. We had the machine programmed to build a three level building with living quarters on the lowest (and safest) level, with office, labs, and manufacturing on the upper levels. There was even a garden area to grow vegetables to enhance our diet and produce some oxygen and filter out some carbon dioxide.

We turned on the machine and it started printing the bottom level of the building. All we had to do was keep feeding it the raw lunar regolith materials, the silicate binder, and electrical and piping which it would place and hold in locations as it poured walls and those materials became locked in place by the hardening walls. We wanted more solid materials like rebar to support the structure but didn't have the capacity to carry those materials to the Moon. Instead we were counting on the walls to be hard enough and carry enough load to make the structure solid.

Our building would be like those built by the Romans—who invented concrete. The Romans would use volcanic ash and lime in their concrete which they used to build many seaports and famous building like the Roman Colosseum. The Romans also didn't use rebar and most of their concrete structure were pure concrete like the Pantheon which was built with different thicknesses of concrete and well thought out geometric designs to give it the strength to hold up for two thousand years.

Over the next several days the building machine first printed a floor for the whole structure, then we could see the walls rising on the basement level which was designed to be twelve feet tall to give a sense of space. It would also have a hanging ceiling with air and other utilities in it which would make the visible height ten feet.

The plan was to finish and roof over the basement with its own airlock entrance as the rest of the building was completed. An elevator shaft was installed but blocked off temporarily. A ramp actually led up to the airlock on the first floor.

As the first level was completed Olga and I started moving in environmental systems and connecting them up. First was the air generation and ventilating system. It worked off of water ice which was now being produced from the ice mine. An automated supply ship also landed on Day 25. It homed in on a beacon our people had planted several hundred feet from our main base. The supply ship contained the rest of the initial ice mining and electrolysis equipment to produce usable quantities of fuel and other components we needed like air.

We also hooked up the waste recycling system. This system would dry out human waste and recycle the liquids. The dried waste could be used as fertilizer in the garden. It would be nice to use a toilet again rather than the tubes and bodily waste connections in the temporary inflatable shelter. We had to cut down the percentage of oxygen in the air to reduce fire risks and had brought tanks of liquid nitrogen which we installed with the air equipment to reduce the oxygen percentage to only twenty percent.

Each day Olga and I would take turns so that one of us was monitoring the construction machine while the other was working on systems in the building basement. After two more weeks the basement had a heater installed and air working inside. It also had a six inch thick ceiling to keep out solar radiation.

This was fortunate because the Commander called us all together early before dinner and we wondered what was up. He told us Earth was advising us of a large solar storm which would hit within six hours. It was projected to last several days and the inflatable tent would not provide enough protection. We would basically be barbequed if we stayed in the tent.

The main option was for us to live in the lander which was so compact nobody really wanted to do it. Both Olga and I suggested the basement of our structure was ready for occupation and would be safer than the lander because of its six inch thick ceiling. It would be ideal for a larger solar flare to have our full three foot thick roof. But given the projections of the flare, the current roof should work. Also due to the Moon's angle to the Sun and orbit around the Earth, we wouldn't be exposed to radiation problems for most of the storm. The Commander asked us more questions to be sure of the safety but we could see the relief on his face that we would not need to go live in the lander again for days.

Pretty soon he was assigning everyone tasks to move our food, sleeping equipment and more to the main shelter's basement. Over the next few hours we were a beehive of activity as we moved everything we could into the basement of the in progress building. That night we were

all in the unfinished basement and it was pretty messy, but all of our life support systems were working properly.
We had radiation monitors all over the basement and the only area which registered dangerous was out next to the airlock upstairs.

We spent the next several days playing cards and watching movies while waiting for the solar storm to finish. The next morning we were given the all clear by Earth and resumed construction. Now you could see the outline of the walls on the main level as the printed building continued to grow.

15.3 Finishing the Shelter

Olga and I restarted construction on the building that day. She worked outside while I worked on the interior of the basement level. This included setting up partitions for rooms. The pre-fab partitions were constructed outside by the construction machine. Then I would take them inside and position them for the rooms. They would then snap together to form walls and even doorways. The walls would fit in slots in the floor which were part of the original construction. I only had to do some drilling and screwing to connected power, doors, and more for each room.

As the rooms were built everyone started moving their personal items into them. Then I also started working on the kitchen and galley area. I got some help from other crew members who wanted this finished as quickly as possible.

Going outside that afternoon I could see that the walls were rising well on the main floor which contained some pre-defined rooms. The rest of the rooms would be based on movable partitions.

Away from the main shelter construction continued. The ice mining operation and splitting into component oxygen and hydrogen was now looking pretty close to completion. Over the next few weeks the main floor of the shelter was finished and the smaller third floor was now under construction. You could see the main roof taking shape as the third floor grew. After another week of construction the third floor overall structure and roof was completed. The structure had no windows to keep the interior radiation safe. Windows would be simulated from exterior cameras which could display on large window screens inside.

My next big task was to run the tractor to push regolith over the roof. The idea was to bury it with at least several feet of covering to provide full protection well beyond any type of projected solar flare we could imagine.
I started by building a ramp on the side of the shelter away from the two airlocks. It took me most of that day to build the ramp. The next week was all about plowing regolith onto the roof. The regolith was then compacted in place by the building machine.

Finally, after over a month of construction you could look at the full outline of main shelter. It didn't look like much from the outside. All you could see from there was a big pile of soil with an entrance ramp and ramp cover going into a dark cavity. There were actually airlock ramps on two sides of the structure. In case one was needed as an emergency entrance.

When you entered the airlock you waited for the air pressure to be equalized. Then the door would open and you would enter into the equipment room with racks for spacesuits, and lockers with other outside equipment. Then a pressure tight door opened into the main shelter. At this location there were stairs up to the second level or down into the basement. A freight elevator was also next to the stairs to take larger equipment up and down. While the basement living area was pretty much finished, there was still a lot of construction on levels one and two. This construction would go on for months and the next crew rotation would also be continuing building.

All About Moon Bases-And Our Plans to Return to the Moon

16.0 Types of People Needed

The people needed to live and work at a Moon Base will have different backgrounds from our current Astronaut population. They will all probably be in their thirties and forties with previous experience in their careers.

Stays at the Moon Base will be similar to those at the Antarctic South Pole Base or International Space Station then get longer as capabilities of the base increase. One thing our new Moon visitors will probably not need to worry about is bone deterioration. One sixth gravity is light but it is a much healthier environment than microgravity. Still medical research on the long term effects of living on the Moon will need to be conducted.

Here are some of the potential types of work and careers for working and living on the Moon:

Pilots

NASA pilots will of course still be needed for the Orion Spacecraft, the Lunar Lander, and other equipment which needs guidance and control

Engineers

Engineers of all types will be needed to manage and repair high tech systems. I'm sure that there will also be many experiments which need a lot of in depth technical understanding to run.

Geologists

Understanding the Moon's different craters, Maria, and layering requires knowledge of how these things might be formed. Also to find the best locations to mine water ice at the South or North Pole.

Miners

Mining Water Ice is going to be very important for potential rocket fuel and water for the residents. Knowing mining techniques and how to use mining equipment will be very important.

Agronomists/Farmers

It's too expensive to transport all the food from the Earth to the Moon especially for long term residents. The ability to grow fruits and vegetables using hydroponics and lunar regolith for soil will help keep the costs down and offer a nice variety and fresh food for lunar residents.

Astronomers

With the Moon not having any atmosphere it will be a perfect location for lunar observatories. Astronomers would really enjoy the Moon for telescope locations.

Systems Engineers

There will be lots of data and communications networks at the Lunar Gateway and Moon Base and will need technical persons to setup, use, and maintain them. Lots of other high tech equipment will also be used and need in depth technical understanding to deploy, use, and maintain.

Construction Experience

The Moon Base will need initial and continuing building and enhancements. All the same specialty constructions skills used on Earth may also be needed on the Moon. Regular astronauts can perform some of the initial tasks which are well pre-defined, but eventually persons with the actual construction skills will be need for base maintenance and expansion.

Medical Doctors/Nurses

We will also need medical professionals to staff a Base since other medical treatment options on Earth will be days or even over a week away.

All About Moon Bases-And Our Plans to Return to the Moon

17.0 Working at the Moon Base

What are the reasons to build and support a Moon Base? In this chapter we provide some reasons.

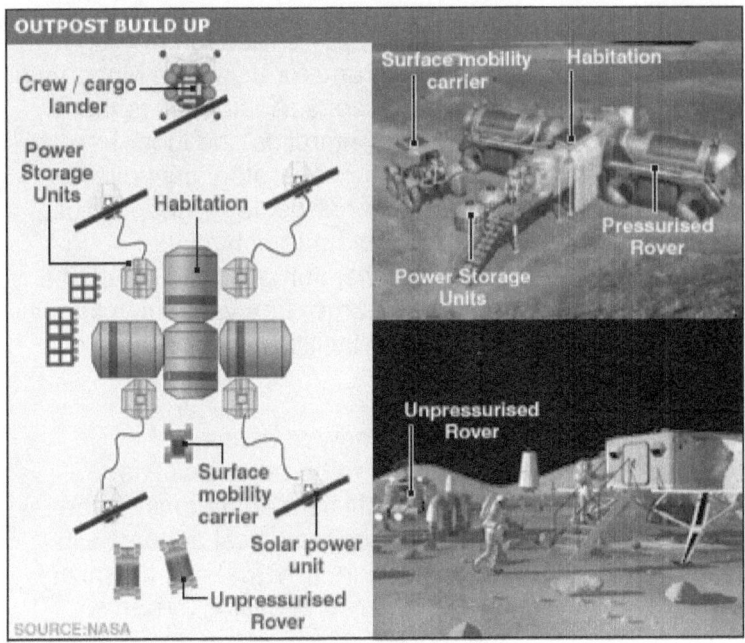

17.1 A Dry run for Visiting Mars

NASA's stated purpose for the Artemus program is to place people on the Moon's surface and develop an ongoing presence there. This is all being done to put more infrastructure and plans in place for launching a manned mission to Mars.

So one of the main missions of a Moon base is to learn how to live, work, study, and generally learn how to exploit the resources of the Moon.

17.2 Scientific Research

A lunar base will create new opportunities for investigating the Moon and its environment and for using the Moon as a platform for scientific investigations. Analogous to the function of McMurdo Base in Antarctica, the lunar base will provide logistical and supporting laboratory capability to rapidly expand knowledge of lunar geology, geophysics, environmental science, and resource potential through wide-ranging field investigations, sampling, and placement of instrumentation. Access to large, free vacuum volumes may enable new experimental facilities such as macro particle accelerators.

The fixed platform will enable new astronomical interferometric measurements to be obtained. The challenge of long-term, self-sufficient operations on the Moon can spur scientific and technological advances in materials science, bioprocessing, physics, and chemistry based on lunar materials, and reprocessing systems.

17.3 Exploitation of Lunar Resources

It has been argued that major industrialization of space cannot occur without access to the resources of the Moon. This might include immense projects such as solar power satellites.

A radio telescope located on the far side of the Moon would be shielded from background noise generated by terrestrial sources. An initial lunar instrument may well be a phased array of dipoles to be demonstrated at a sufficiently large scale. It is reasonable to develop the resource potential of the Moon to offset the high Earth-to-orbit transportation costs (Hearth, 1976). The lower gravitational field of the Moon and the absence of an atmosphere that retards objects accelerated from the surface provides a potential 20- to 30-fold advantage for launching from the Moon instead of Earth. For example, at liftoff, about 1.5%of the space shuttle's mass is payload. Most of the mass is propellant. From the Moon, approximately 50%of the mass can be payload.

The commodity currently envisioned to be most in demand in Earth-Moon space over the next three decades is liquid oxygen, which makes up 6/7 of the mass of propellant utilized by cryogenic (hydrogen-oxygen) rockets, such as the Centaur or postulated ones. Although it would appear unlikely that an atmosphere less body is a source for oxygen, it is actually an abundant element on the Moon (Arnold and Duke, 1978). It must be extracted, however, from silicate and oxide minerals into its liquid form for use as a propellant. Several processes have been suggested (Criswell, 1980) for accomplishing this, including reduction of raw soil by fluorine (which is recovered) or reduction of iron-titanium oxide (ilmenite) by hydrogen (also recovered). Preliminary laboratory studies have verified the concepts behind some of these processes.

Systems studies (e.g.,Carroll et al., 1983) show that oxygen production on the Moon could benefit STS in the early years of the next century, even if the hydrogen component of the propellant needed to be brought from Earth water at the lunar poles (Arnold, 1979) or extracting the dispersed solar wind-derived hydrogen in the lunar regolith would greatly improve the economics of the transportation system.

Other commodities also could be produced Metals, such as iron or titanium, can be extracted from the lunar soil or from specific rocks or minerals with differing degrees of difficulty. For example, small quantities of metal (primarily iron) from meteorites can be concentrated with a magnetic device from large amounts of lunar soil, or, with much larger energy inputs, titanium can be obtained from ilmenite. These products could find applications in large space structures. Lunar Titania or alumina might be used to produce aero brakes (heat shields). In the long term, at relatively high levels of development, production of components for solar electric power generation in space (e.g., solar power satellites) could be made feasible (Bock, 1979).

18.0 Moon Base Future Growth

Once a Moon Base is established the plan is for the settlement to grow and expand to support research and other Moon related tasks.

Some plans have been made in the 1960s and 1970s for building a base and its plan for growth. Here are the possible stages from one report:

Phase I: Preparatory exploration

- Lunar orbiter explorer and mapper
- Instrument and experiment definition
- Site selection
- Automated site preparation

Phase II: Research outpost

- Minimum base, temporarily occupied, totally resupplied from Earth
- Small telescope/Geoscience module
- Short range science sorties
- Instrument package emplacement

Phase III: Operational base

- Permanently occupied facility
- Consumable production/Recycling pilot plant
- Longer range science sorties
- Geoscience and Medical laboratories
- Experimental lunar radio telescope
- Extended surface science experiment packages

Phase IV Advanced base

- Advanced consumable production
- Satellite outposts
- Advanced geoscience laboratory
- Plant research laboratory
- Advanced astronomical observatory
- Long-range surface exploration

Phase V: Self-sufficient colony

- Full-scale production of exportable oxygen
- Volatile production for agriculture, Moon-orbit transportation
- Closed ecological life support system
- Lunar manufacturing facility: tools, containment systems, fabricated assemblies, etc.
- Lunar power station-100% lunar materials-derived
- Expanding population base

Looking to the future

Below are some drawings for a future expansion of Moon Bases into a full city with a lot of it built underground.

Long term the Moon offers a good base for exploration of the Solar System. With its one sixth of an Earth gravity many things can land there for refueling and maintenance.

Also because of the lower gravity well, a linear accelerator could be built there to launch supplies and manned ships into orbit.

The Moon also offers options to give humanity another location to live in case of worldwide disasters on Earth.

19.0 Summary

Humanity will go back to the Moon and build a Moon Base. President Trump has speeded up the prior plans so we now have a return Moon landing set for 2024.

Lots of new technologies will be used to build the new Lunar Gateway and for the eventual Moon Base.

The technologies developed for the International Space Station will also help provide systems we can also use on the Moon Base.

Building a manned Lunar Base is a necessary step for a manned Mars mission and to send man to other objects in the Solar System.

I hope this book helps encourage many of you who are younger to participate in this exploration of our Solar System.

Martin K. Ettington

December 2019

All About Moon Bases-And Our Plans to Return to the Moon

20.0 Bibliography

1. Colonization of the Moon.
https://en.wikipedia.org/wiki/Colonization_of_the_Moon.
[Online]

2. Moon EXploration Infographic.
https://www.space.com/26541-moon-exploration-350-year-history-infographic.html. [Online]

3. 2001 Movie Moon Base.
https://2001.fandom.com/wiki/Clavius_Base. [Online]

4. The Lunex Proposal.
https://en.wikipedia.org/wiki/Lunar_outpost_(NASA). [Online]

5. Project Horizon.
https://en.wikipedia.org/wiki/Project_Horizon. [Online]

6. Lunar Outpost.
https://en.wikipedia.org/wiki/Lunar_outpost_(NASA). [Online]

7. Shakelton Crater.
https://en.wikipedia.org/wiki/Shackleton_(crater). [Online]

8. Interesting Moon Base Proposal.
https://interestingengineering.com/8-interesting-moon-base-proposals-every-space-enthusiast-should-see. [Online]

9. 5 Fascinating Things You may not have known about the Moon. *https://www.techeblog.com/5-fascinating-things-that-you-may-not-have-known-about-the-moon/.* [Online]

10. Ambitious Plans to Colonize the Moon. *https://www.mentalfloss.com/article/53588/15-ambitious-plans-colonize-moon.* [Online]

11. Lunar Lava Tubes. *https://en.wikipedia.org/wiki/Lunar_lava_tube.* [Online]

12. Nasa secret plans for a Moon Base. *https://www.foxnews.com/science/secret-nasa-plans-for-moon-base-and-37-rocket-launches-revealed.* [Online]

13. *Lunar Base Concepts.* Institute, The Lunar and Planetary.

14. LESA Moon Base. *http://www.astronautix.com/l/lesalunarbase.html.* [Online]

15. NASA Lunder Lander Requirements. *https://spaceflightnow.com/2019/10/07/nasa-opens-competition-to-build-human-rated-lunar-landers/.* [Online]

16. How To Build a Moon Base. *https://www.nature.com/articles/d41586-018-07107-4.* [Online]

17. Cost to Live on the Moon. *http://www.astronomy.com/news/2016/09/how-much-it-would-cost-to-live-on-the-moon-in-9-minutes.* [Online]

18. Lunar Power Generation Study.
https://www.thespacereview.com/article/2882/1. [Online]

19. Lunar Water Mining.
https://www.nasa.gov/directorates/spacetech/niac/2019_Phas e_I_Phase_II/Lunar_Polar_Propellant_Mining_Outpost/. [Online]

20. Lunar Exploration Partners. *https://spacenews.com/nasa-sees-strong-international-interest-in-lunar-exploration-plans/.* [Online]

All About Moon Bases-And Our Plans to Return to the Moon

All About Moon Bases-And Our Plans to Return to the Moon